The Structure of Design

The Structure of Design

An Engineer's Extraordinary Life in Architecture

Leslie Earl Robertson

Edited by Janet Adams Strong

THE MONACELLI PRESS

Library of Congress Cataloging-in-Publication Data

Names: Robertson, Leslie E., author.
Title: The structure of design : an engineer's extraordinary life in
architecture / Leslie Robertson.
Description: First American edition. | New York : The Monacelli Press, 2016.
Identifiers: LCCN 2016022306 | ISBN 9781580934299 (hardback)
Subjects: LCSH: Robertson, Leslie E. | Engineers--United States--Biography. |
Structural engineering. | Architects and engineers. | BISAC: ARCHITECTURE
/ Individual Architects & Firms / Monographs. | BIOGRAPHY & AUTOBIOGRAPHY
/ Artists, Architects, Photographers. | TECHNOLOGY & ENGINEERING /
Construction / General.
Classification: LCC TA140.R55 A3 2016 | DDC 624.1092 [B] --dc23
LC record available at https://lccn.loc.gov/2016022306

Design: Chris Grimley, Shannon McLean and Brett Pierson. over,under

ISBN 978-1-58093-429-9

10 9 8 7 6 5 4 3 2 1

Printed in Singapore

www.monacellipress.com

Contents

III Collaborations with Architects 97

DEDICATION

I owe everything to the extraordinarily delightful, and wonderful, Saw-Teen See: my companion, my lover, my wife, my compass, my challenger, who continues to change me for the better, breathing life into my body and my soul, while bringing happiness and love into every facet of the existence of this fortunate engineer.

ACKNOWLEDGMENTS

The making of wonderful projects is not the creation of the he or the she whose name is on the door. It is truly a collaborative effort, with every individual involved providing some level of participation. The good, the bad, and the indifferent are the products of us all.

This book was written with the able assistance of so many persons as to leave me bewildered at the depth of their stalwart support. The horizons of my career were created by countless architects, artists, engineers and friends who took the time to share with me their knowledge, their philosophies, and their experiences. Where not found in this book, I cannot even begin to list the remainder here.

Ms. SawTeen See, brilliant engineer, wife and compatriot, put her shoulder to the wheel and inspired me to diligently sustain the effort that went into this book. Beyond reviewing the writing, she delved into depthless files to obtain information, photographs, images, data, and the like to make more interesting and readable this book, and both awoke me in the morning to get to work, and then pushed me away from the writing in the early evening to maintain my sanity. Without her I would not have set out on the adventure of this book, nor would I have been able to finish it.

Without the guiding hand of Ms. Janet Adams Strong, the completion of this book would not have been possible. She prevailed in her boundless determination to improve my writing and correct my "unusual" syntax and imaginative spelling (at least for the most part). I believe that she exerted more energy in assisting me in the completion of my writing of this book than she had on any other of her countless projects. For her efforts in the long and intensive process of completing this book, I remain forever grateful.

Alan Rapp, Senior Editor at The Monacelli Press, while providing invaluable guidance in so many ways, remained ever diligent, ever perceptive, toward making this book more interesting and more readable. The very talented book designer Chris Grimley overcame uncountable obstacles to produce the beautiful document now in your hands.

PREFACE

I am an engineer specializing in architectural structures; many think of me as the engineer of the two towers of the World Trade Center, or of the Miho Museum Bridge, but there are many other buildings and structures, both famous and lesser-known. I have been extraordinarily fortunate in having the opportunity to work with and to become friends with some of the most wonderful and famous architects of our time, as well as with their staffs. This book recounts some of the professional and personal events that have filled my life.

While a host of friends and compatriots had urged me to write a book, this book, I resisted . . . and resisted. One of those most insistent, the late Curtis Bill Pepper—a skilled author, journalist, and dear friend—convinced me that I really should and could write a book about my career as an engineer in architecture. I coerced myself into believing that I had a message to tell and that, just perhaps, that message could be partially expressed in the relating of my experiences with architects and with those projects that were produced, at least in part, under my baton.

Taking Bill's encouragement to heart, but needing time and a clear mind to organize my thoughts. I travelled to the MacDowell Colony, an extraordinary artist retreat in New Hampshire where I had been fortunate in having been admitted on a prior occasion. There, completely isolated from the trivia that occupies much of our lives, I could devote all of my energies toward the thinking and the writing.

After unpacking and setting up my computer, with thoughts of Bill filling my mind, I asked myself if I should pursue a sort of architectural Moby-Dick, a tale of obsession, though, unlike Melville's story, not fiction and not of death. Could it be narrated as though I had watched the World Trade Center come crumbling down, dashing my dreams of the Towers as a great spirit that my mind had sought to realize in architecture, defying gravity, time, and space? My tale could somehow project the emotional upheaval within me as the demise of the World Trade Center, one of the foundations of my life's work, wrenched at my being, compounded by the knowledge that, perhaps, I could have made the towers stronger, more stalwart? Or could my story be narrated through a more realistic lens, with SawTeen entering my life . . .

first as a professional collaborator, later to become the cornerstone of my being? How would I weave into this my passionate advocacy for civil rights, women's rights, and world peace, including the people and projects that I have interacted with in one way or another? Could it be not just a conventional compendium of my work, but somehow expressing something more personal: a spirit and a mind seeking to realize itself in architecture. I struggled those first hours with my own perceived limitations and insecurities. The skills of this engineer did not seem to easily or immediately adapt to that which it takes to write a book. What was I really able to do?

I left these concerns behind in my studio in the woods to join the other dozen or so MacDowell Fellows for dinner, where I learned of many of the inspiring activities of these talented spirits: artists, composers, and writers. Following dinner, a delightful photographer, Ms. Kelli Connell, provided a glimpse of her incredibly imaginative and thoughtful work. Inexplicably, this encounter with Kelli and her work helped me reclaim my confidence; I retired to my studio, convinced that I could do it, that I could write a book that recounted the tales of my interactions with some of the wonderful architects who have allowed me to work with them and with some of the astonishing projects that we were able to create together. On that joyous and wonderful evening at MacDowell, I set out to write this book.

Since over my life I have not kept a diary, much to be found herein is taken from memory and from such scraps of thoughts that I had penned lo those many years ago. Without question, as with almost everyone, I have a memory that is both fallible and convenient. It is not my intent to document a mountain of inert facts and figures. Instead, I hope that some of the passions in my heart are able to be discerned from the reading.

Above all, I hope that, from this book, young architects and engineers will learn from my life-experience something of the joys as well as the trials and the tribulations of our professions. Indeed, it is exactly this point that has generated within me the energy required to bring this book to you.

I

Selected Projects

This illustrated overview is intended to provide a brief glimpse of the scope of the work accomplished over the course of my career, encompassing projects that I, or SawTeen See and I, participated in the leadership of the engineering efforts. Far more projects are unlisted, including those that remain confidential, as well as most peer reviews and competition entries. Incorporated are only those projects, both large and small, that I believe to be relevant to the thrust of this book.

Given the variances in circumstances and the national laws and affordances particular to each project, our roles have varied widely. We have provided concept design, design development, tender phase, construction documents, construction phase, and more. For some projects, beyond idea generation, little more is required of us; for others, our services may carry straight through to completion. While there are notable exceptions, in the US our client is usually the architect; in other countries our client is commonly the owner or the developer.

The date given for each project is usually the date of an established agreement with the client or of the initiation of our work. While we recognize that design and construction processes are accomplished by coordinated teams, space allows only listing the primary architectural collaborators; the name of the principal architect or designer who collaborated with us appears in parentheses after the name of his or her office. Projects with their own entries or extended mention later in the book are indicated with corresponding page numbers in parentheses above the photograph.

1956 (p. 61)

Rincon Island and Bridge
Rincon Island, California
John A. Blume Associates

A 2,700-foot all-concrete bridge, with single piles alternating with pairs of battered piles, required 25% less piling than that of conventional designs, while reducing construction time and cost. The island, newly ringed with tetrapods to resist oncoming waves, protected a host of oil wells that were slant-drilled.

1960 (p. 62)

General Rafael Urdaneta Bridge
Maracaibo, Venezuela
Raymond International

An offshore storage platform owned by Raymond International had collapsed; careful work uncovered defects in the design. An alternative solution was developed for the foundations of this concrete bridge, making use of meter-deep steel wide-flange shapes in lieu of conventional reinforcing steel.

1960 (p. 64)

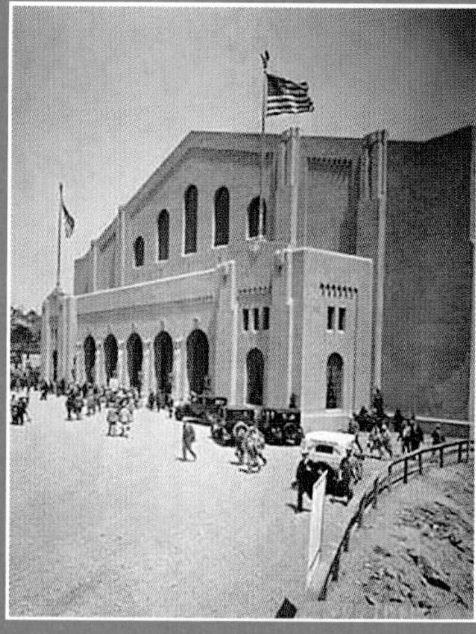

Convention Center and Opera House
Seattle, Washington
James J. Chiarelli, Architect
(B. Marcus Pretica)

A flat-floored building was transformed into a 3,100-seat, balconied theater. The main floor became a parabolic bowl, the stage was deepened, and the gridiron was enlarged and raised. Extensive post-tensioning reduced the weight of the new construction, limiting the need for new piles.

1961 (p. 65)

IBM Building
Pittsburgh, Pennsylvania
Curtis & Davis
(Nathaniel C. Curtis)

The first known building to use quenched and tempered steels; the largest piece in the perimeter is a 6-inch angle. Also innovative was the prefabrication of the trusswork. More than a half century later, this same diamond-shaped truss is marketed by some as a new concept: the "diagrid."

1961 (p. 103)

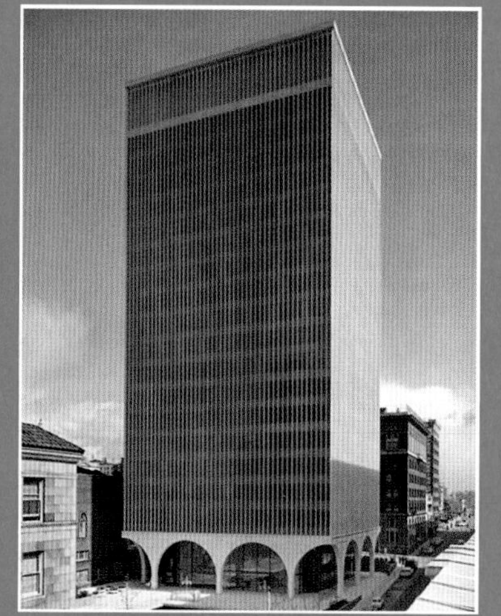

IBM Building
Seattle, Washington
Minoru Yamasaki & Associates
(Minoru Yamasaki)

Twenty-two stories high and with concrete floors, the perimeter columns consist of 4-inch pipes. Quenched and tempered steels were used toward the base while, in the middle, combinations of yield point and wall thickness were employed. Nearing the top, common steels were used.

1962

North Shore Congregation Israel
Glencoe, Illinois
Minoru Yamasaki & Associates
(Minoru Yamasaki)

Fourteen concrete shells forming the roof are delicately connected to enhance the architectural design while providing wind and earthquake resistance. The shells were sufficiently identical to enable the reuse of only one form. The decorative concrete perimeter walls are precast concrete.

1963

(p. 108)

World Trade Center
New York, New York
Minoru Yamasaki & Associates, Emery Roth & Sons
(Minoru Yamasaki, Aaron Schreier, Julian Roth)

Bombed in 1993 and then destroyed on September 11, 2001, structural innovations for the towers were numerous, including the first tubular framing system to be used at this scale and cutting-edge wind engineering developed with Alan G. Davenport. The information required for the detailing and the fabrication of the structural steel was provided in a digital format with IBM punched cards. Steel production and/or fabrication was sourced in Canada, Europe, Japan, and across the US. One steel detailer was used for the more than twenty fabricators.

1963

Japanese Cultural Center
San Francisco, California
Minoru Yamasaki & Associates
(Minoru Yamasaki)

Nothing was standard in this complex multiuse project, which includes a hotel, theater, restaurants, stores, exhibition space, a pedestrian bridge, and a concrete pagoda. Our designs were reviewed with suspicion by the Building Department but were ultimately accepted.

1967

Manufacturers & Traders Trust
Buffalo, New York
Minoru Yamasaki & Associates
(Minoru Yamasaki)

Housing the headquarters of a major bank in the deteriorated downtown core, this building was an urban catalyst with a 70-feet clear span for office flexibility. The base arches and the closely spaced columns are a signature of a Yamasaki design.

1965

(p.173)

US Steel Tower
Pittsburgh, Pennsylvania
Harrison, Abramovitz & Abbe
(Max Abramovitz, Charles Abbe)

With each of its 64 floors one acre in size, this was the largest privately owned building in the world upon completion in 1970. The structural system successfully demonstrates the possibilities of structural steel. Borrowing from the World Trade Center, the structure makes use of outrigger trusses to stiffen the building and to moderate the thermally induced expansion and contraction of the external columns. We provided the basic plan, form, and spacing of the liquid-filled perimeter columns. A thoroughly "modern" building, the roof was designed for vertical-takeoff jet aircraft.

1968

(p.193)

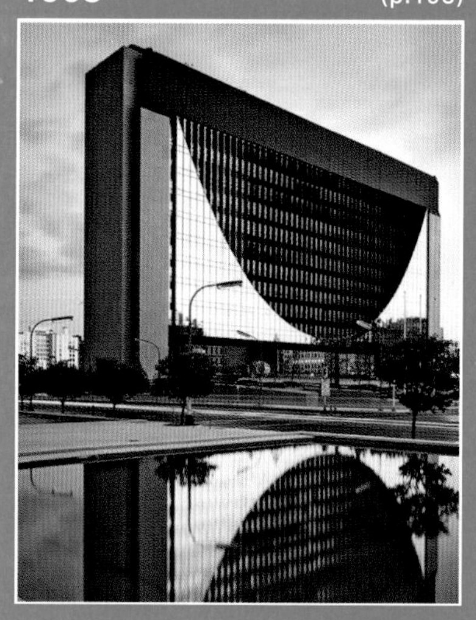

Federal Reserve Bank
Minneapolis, Minnesota
Gunnar Birkerts and Associates
(Gunnar Birkerts)

This is the first major building to be carried on post-tensioned catenary arches. Simplifying Birkerts' idea to use multiple catenaries, we introduced a single catenary. Structural designs provided for a future 50 percent vertical expansion.

1969

IBM Sterling Forest
Sterling Forest, New York
Gunnar Birkerts and Associates
(Gunnar Birkerts)

Birkerts took advantage of a gardenlike site to create a handsome data center and office building, constrained by a modest construction budget and tight design schedule. The all-steel post-and-beam structure could have been designed on the back of an envelope but required computer assistance.

1969

Condon Hall, University of Washington
Seattle, Washington
Mitchell/Giurgola Architects
(Romaldo Giurgola)

While, sadly, Romaldo Giurgola passed away in 2016, the design of this cast-in-place concrete building remains as a part of his legacy. Housing classrooms and a library, the design shields direct sun while maximizing natural light inside.

1969

(p.321)

Greenacre Park
New York, New York
Sasaki, Dawson, DeMay Associates
(Masao Kinoshita)

This is a lovely vest-pocket park in dense midtown Manhattan. David Rockefeller participated in the most of the project meetings. While anticipating that slabs would be soil-founded, drainage and plumbing systems required structural slabs. Adjacent buildings required underpinning.

1969

Century City Plaza
Los Angeles, California
Minoru Yamasaki & Associates
(Minoru Yamasaki)

This complex includes twin 500-foot triangular office towers, a hotel, exhibition hall and athletic facilities, and a 6,000-car underground garage. All of the perimeter columns of the office towers are carried to the three corners by the 250-foot span Vierendeel trusses of the perimeter frame.

1969

I-Beam Sculpture 1
Broome County Government Plaza
Binghamton, New York
Masao Kinoshita Studio
(Masao Kinoshita, sculptor)

Common weathering-steel shapes were organized with thoughtfulness and care, allowing Kinoshita to create a light and airy sculptural form. Ignoring the piece-by-piece construction schedule provided, the contractor simply shored the lot.

1970

Penn Mutual Life Insurance Company
Philadelphia, Pennsylvania
Mitchell/Giurgola Architects
(Romaldo Giurgola)

The 90-foot clear span from the outside wall to the services core was a major accomplishment for its time. The architects inserted an existing facade of Egyptian styling and a partial mezzanine, thus hiding the great lobby space created by that huge span.

1970

St. Bede Abbey Worship Assembly
Peru, Illinois
Mitchell/Giurgola Architects
(Romaldo Giurgola)

The architectural design required extensive remodeling, including the structural design of very large interior doors clad in leather. Our engineering role was largely limited to the nonstructural components.

1970

Temple Beth El
Bloomfield Township, Michigan
Minoru Yamasaki & Associates
(Minoru Yamasaki)
The structure of this steel and wood temple is a forerunner of Yamasaki's design for the much larger, all-steel Meishusama Hall, Shigaraki Prefecture, Japan.

1970

(p.165)

Meishusama Hall
Shigaraki Prefecture, Japan
Minoru Yamasaki Associates
(Minoru Yamasaki)
Yamasaki and structural engineer Yoshikatsu Tsuboi came to an impasse over the skylights, the design of which impacted the earthquake resistance of the hall. To circumvent the controversy, a third design was developed to the satisfaction of both.

1971

INA Building
Philadelphia, Pennsylvania
Mitchell/Giurgola Architects
(Romaldo Giurgola)
Penthouses were expanded and reshaped to protect against unwanted noise. Tender documents for structure were completed before the architects had finalized their schematic designs, the result of an in-depth understanding of and appreciation for the style of the architectural designers.

1973

Municipal Fire Headquarters
Corning, New York
Gunnar Birkerts and Associates
(Gunnar Birkerts)
A simple and economical structural system was devised for this garage and office building with sleeping facilities. With a low budget, the building was quick to construct, providing spaces that were efficient for people and ample parking for fire trucks.

1973

Sherman Fairchild Center for
Life Sciences, Columbia University
New York, New York
Mitchell/Giurgola Architects
(Romaldo Giurgola)
Floors were added atop an existing, occupied, aging laboratory building that was not originally designed for the loads from the added floors. Challenges included construction-generated vibrations that affected precision instruments.

1973

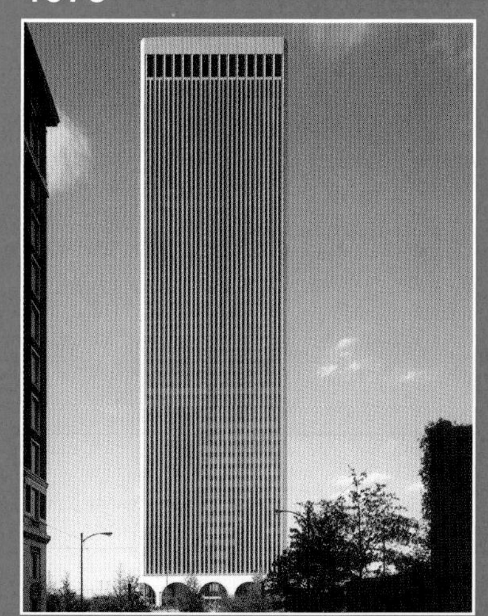

National Bank of Oklahoma
Tulsa, Oklahoma
Minoru Yamasaki & Associates
(Minoru Yamasaki)
Design was relatively straightforward up to the point that it was discovered that the contractor had mistakenly omitted half of the reinforcing steel from the second-floor slabs even as construction continued upward. To correct the contractor's error, reliable repairs were implemented.

1971

Prudential Insurance Building Co.
Roseland, New Jersey
Grad Associates
(Harry Mahler)

This office building uses 90-foot span spandrel girders but accommodates the differing needs for vertical deflection control between the spandrels and that of the curtain wall. Computer analyses revealed that the testing laboratory had fabricated strength data for the concrete test cylinders.

1971

(p.197)

IBM Manufacturing Plant Repairs
Sumaré, Brazil
IBM (client)

The roof of this IBM manufacturing facility sagged excessively in rainstorms, indicating a design or construction flaw. After multiple investigatory trips to Brazil, repairs were implemented to correct deficiencies in thousands of connections and members.

1971

Fisher Brothers Bridge
New York, New York
Minoru Yamasaki & Associates
(Minoru Yamasaki)

The first post-tensioned concrete bridge in New York, which spanned southerly over the street from The World Trade Center. From bridges designed by John V. (Jack) Christiansen from our offices, we borrowed the concept of a hyperbolic paraboloid soffit to accomplish the changes in depth.

1974

Calvary Baptist Church
Detroit, Michigan
Gunnar Birkerts and Associates
(Gunnar Birkerts)

The design of this new building for an historic black church is innovative and exciting, including a huge mirror behind the pulpit that reflects the congregation's vitality. Cast-in-situ walls composite with metal siding were designed for the ground-level areas.

1974

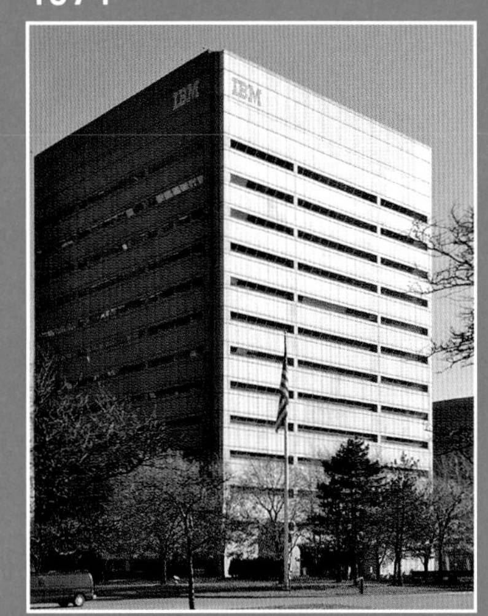

IBM Office Building
Southfield, Michigan
Gunnar Birkerts and Associates
(Gunnar Birkerts)

This 14-story building on a prominent site required an evaluation of potential danger in the event of a partial or total collapse of an adjacent tall, cable-stayed tower. The innovative facade was environmentally designed to control interior sunlight.

1974

Living History Center
Philadelphia, Pennsylvania
Mitchell/Giurgola Architects
(Romaldo Giurgola)

This is a straightforward museum that included an IMAX theater designed by Raymond Loewy. As it turned out, our client was not a licensed architect, attempting to shift that burden onto us; Loewy's firm went bankrupt within a few years.

1974

General Motors Dual Mode Transit System
Gunnar Birkerts and Associates
(Gunnar Birkerts)

This never-realized theoretical study was designed to expose the advantages and the disadvantages of GM's proposed driverless, computer-controlled transit system. Concept designs were submitted for the guideways, the stations, and for the removal of cars in need of repair.

1974

Century Center
South Bend, Indiana
Johnson Burgee Architects
(Philip Johnson)

Columns in this convention center were spaced around a triangulated courtyard in proportion to their gravity load, with the heaviest-loaded columns closest together.

1975

Strawberry Square, Harristown Key Block I & II
Harrisburg, Pennsylvania
Mitchell/Giurgola Architects
(Romaldo Giurgola)

A multi-building retail and commercial complex located near the State Capitol, with the plan shape driven in part by the city grid.

1975

(p.259)

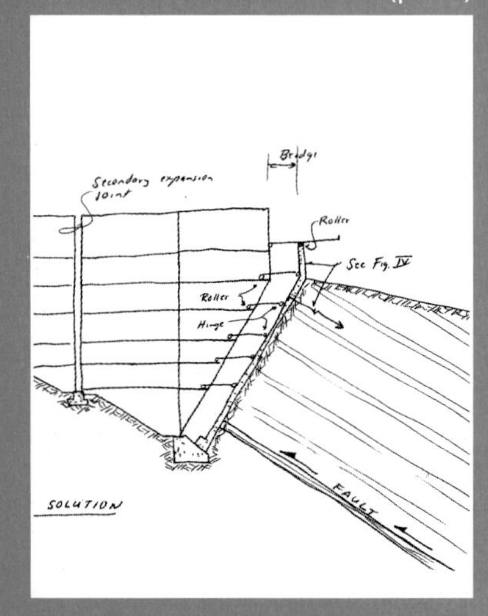

Kapsad Development
Tehran, Iran
I. M. Pei & Partners
(I. M. Pei)

This unbuilt project included a high-rise office tower and surrounding buildings. Once it was discovered that a seismic fault line intersected the project, an unusual, expensive, and likely effective foundation system was developed, but the project was halted nonetheless.

1976

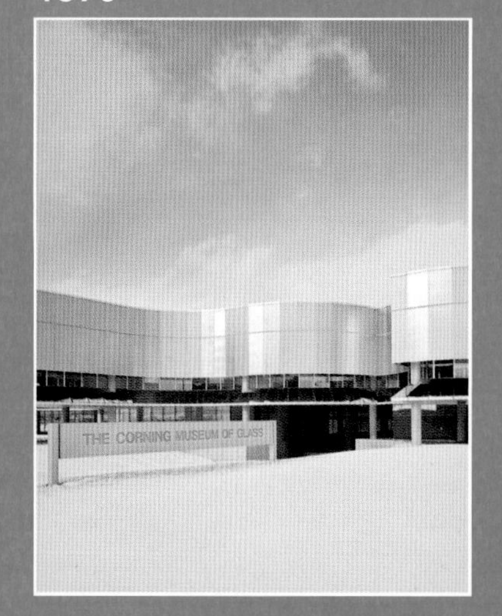

Corning Glass Museum
Corning, New York
Gunnar Birkerts and Associates
(Gunnar Birkerts)

Ingeniously using mirrors tucked under a cantilevering facade, Birkerts developed an environmental design that bounces natural light onto ceilings of the museum. That, coupled with the building's curving form, significantly complicated the structural design.

1976

Howard Savings Bank I & II
Livingston, New Jersey
Grad Associates
(Harry Mahler)

On a suburban corporate-campus office building containing a variety of perimeter facades. Some of the long spans were quite unusual but intrinsic to the architectural image.

1977

Brendan Byrne Arena
Meadowlands, New Jersey
Grad Associates
(Bernard Grad)

The long-span roof is supported by just eight columns. Visually attached to these columns and increasing their apparent bulk are the stairs, elevators, and assorted core building services. The eight columns are located in the plane of the exterior wall; to achieve the cantilevered corners and to create giant frames, column pairs are connected with steel trusses. Similar to the domes of ancient times, a ring of post-tensioning cables bolsters the arching box girders that span to opposing columns. The cables turn diagonally at the corner columns, balancing the loads from opposing columns, while providing the requisite force to resist the thrusting of the interior box girders. The retail and ancillary spaces are created by light framing with the precast seating used as the ceiling above. There is storage for rainwater on the roof.

1976

(p.322)

Waterfall Garden Park
Seattle, Washington
Sasaki, Dawson, DeMay and Associates
(Masao Kinoshita)

The site is underlain with a 40-foot layer of sawdust, which was identified by a long-time neighborhood resident as a former lumber mill. Once confirmed with a geotechnical investigation, the entire park was supported on steel piles.

1976

Desarrollo Morelos Complex
Caracas, Venezuela
Siso, Shaw y Asociados Arquitectos
(Enrique Siso, Daniel Fernández-Shaw)

A complex of high-rise residential and office towers located in an area with a high seismic risk, but with the risk well within that of our prior experience.

1977

Terrace Theater
John F. Kennedy Center
for the Performing Arts
Washington, DC
Johnson Burgee Architects
(Philip Johnson)

For this 500-seat theater, erected atop the roof terrace of the Kennedy Center, Johnson's "classical" design, both in the form and in the structural system, fit well with the history and spirit of Washington DC.

1977

Harbour City Complex
Hong Kong
Eric Cumine Associates
(Eric Cumine)

Limited in height to 200 feet for aviation clearance, these eleven residential and office buildings were the first use of post-tensioned flat plate construction in Hong Kong. Our proposal that the uppermost floor be walk-up allowed the construction of additional floors.

1977

(p.227)

AT&T Headquarters
New York, New York
Johnson Burgee Architects, Simons Architects
(Philip Johnson)

The first building in New York to incorporate steel-plate shear walls stiffened by concrete cladding. These shear walls, while a first for New York, were solidly within our experience. The resulting vertical box girders, creating the elevator and services cores, easily and economically created the lateral force system required for the great height of the floor above the plaza. The stonework of the facade involved more intensive engineering than the entire structural system of the building.

1977

Capital Walkway Bridge
Harrisburg, Pennsylvania
Mitchell/Giurgola Architects
(Romaldo Giurgola)
This is a through-girder steel pedestrian bridge interconnecting the Strawberry Square complex of multi-story buildings.

1977

Neiman Marcus Department Store
San Francisco, California
Johnson Burgee Architects
(Philip Johnson)
Overlooking Union Square and located on the site of the original City of Paris store, the design incorporates replicas of the original oval rotunda and stained glass of the City of Paris store. Structural designs easily obtained required approvals. This project was followed by an addition.

1978

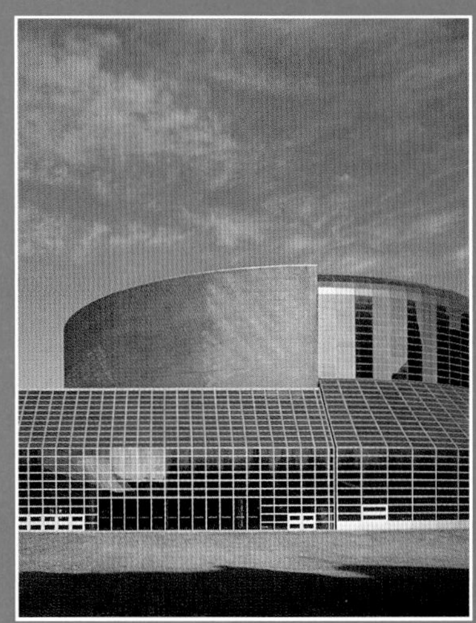

Peoria Civic Center
Peoria, Illinois
Johnson Burgee Architects
(Philip Johnson)
This is an all-steel convention center with an arena, theater, exhibition hall, ballroom, and meeting rooms. The structural complexities associated with the various building shapes in plan, in order to provide handsome structural details, required careful attention.

1978

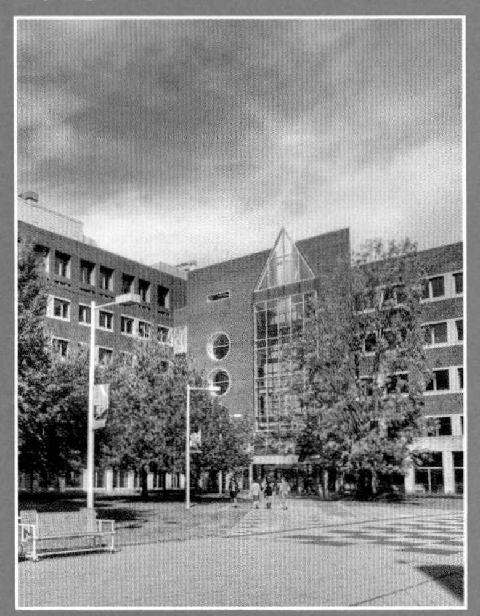

Health Sciences/Services Buildings, MIT
Cambridge, Massachusetts
Gruzen & Partners, Mitchell/Giurgola Architects
(Romaldo Giurgola)
A two-building and atrium complex designed to accommodate a wide variety of occupancies and medical services. The stairs are of intriguing designs.

1978

(p.211)

Citicorp Headquarters
New York, New York
Citicorp (client)
An urgent and clandestine intervention was devised to correct design errors in the structural systems for this 59-story building. The errors could have led to the collapse of the building in a significant windstorm. With admirable directness by the executives of Citicorp, repairs were quickly accomplished.

1978

Charles Shipman Payson Building, Portland Museum of Art
Portland, Maine
I. M. Pei & Partners
(Henry Cobb)
Located on the sites of Cobb's family residences, this museum provides a significant enhancement to the downtown. The structural system, while relatively straightforward, followed rigorously the concept of the skylit architecture.

1978

(p.235)

One PPG Place
Pittsburgh, Pennsylvania
Johnson Burgee Architects
(Philip Johnson)

The complexity of the folded glass curtain wall and the fact that PPG itself had designed the technical aspects of the system, in subcontract to yet another firm, generated extensive complications. In our review of the PPG design, a serious error was discovered in the mullions that had already been fabricated. These elements had to be redesigned, melted down, and re-extruded.

1978

(p.222)

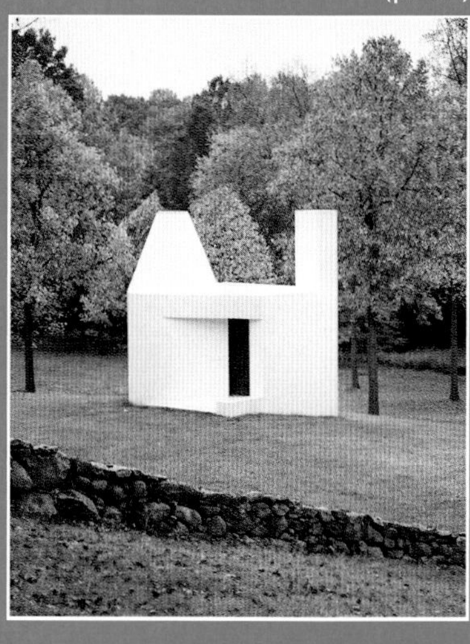

Philip Johnson's Study
New Canaan, Connecticut
Johnson Burgee Architects
(Philip Johnson)

Johnson originally envisioned reinforced concrete for his "monk's cell," but our redesign used less expensive block and brick; the resulting hodgepodge of masonry details were prominently illustrated in the architectural press.

1979

(p.222)

Philip Johnson's Entry Gate
New Canaan, Connecticut
Johnson Burgee Architects
(Philip Johnson)

While Johnson submitted only a crude elevation, our design of the gate followed the concept of a drafter's parallel rule. Johnson initially objected to the gate's capacity to be remotely operable, but ultimately acceded.

1978

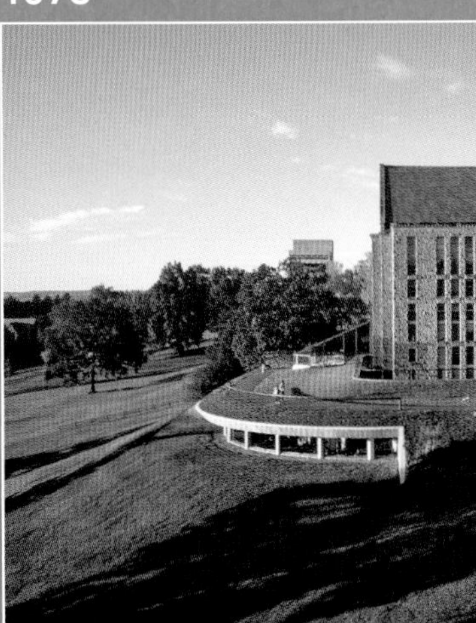

Uris Library Addition,
Cornell University
Ithaca, New York
Gunnar Birkerts and Associates
(Gunnar Birkerts)

This underground building connected to the Uris Library created a wide walking path along the edge of the hill. The design provides extraordinary views over the landscape below.

1980

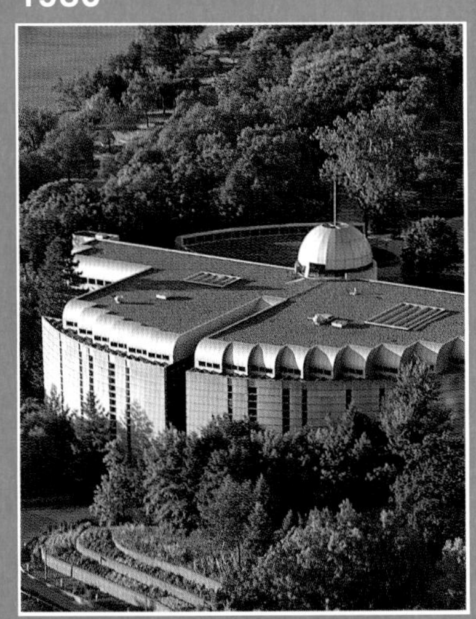

College of Law, University of Iowa
Iowa City, Iowa
Gunnar Birkerts and Associates
(Gunnar Birkerts)

Constructed both on flat land and on a steep hillside, the project had almost every kind of structure: a 1-story concrete portion, a dome, and ever-varying hillside spaces. While the structural system is completely hidden from view, the strength of the building shines through.

1979

(p.207)

Parliament House, Competition
Canberra, Australia
Mitchell/Giurgola Architects
(Romaldo Giurgola)

While our designs contributed modestly to winning the competition, the detailed structural design was left to the local firm of John Fowler. We subsequently produced concept designs for the iconic stainless steel Flag Mast, which is more than 250 feet high.

1980

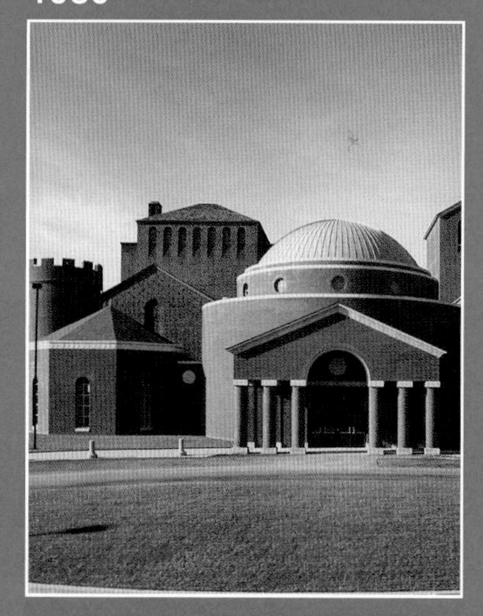

The New Playhouse Theater
Cleveland, Ohio
Johnson Burgee Architects
Collins and Rimmer Architects
(Philip Johnson)
The form was classical, but the structural system had to be thoroughly contemporary, which necessitated a kind of adaptation to a historical mindset to match the character of the architecture.

1980

Weisner Building, Arts & Media Technology Center, MIT
Cambridge, Massachusetts
I. M. Pei & Partners
(I. M. Pei)
Structurally, at least, this is sort of a box-in-a-box. The "architecture" itself was mostly expressed on the perimeter and in the central atrium, while maximizing functionality of interior spaces; the exhibition space is particularly complex.

1980

Pitney Bowes World Headquarters
Stamford, Connecticut
I. M. Pei & Partners
(Henry Cobb)
This low-rise corporate headquarters features complex stairs and atrium bridges. Changes were made in the support details of the stonework, sometimes requiring engineering-driven redesign. The plan shape, responding to a sloping site near the water, is extraordinary.

1980

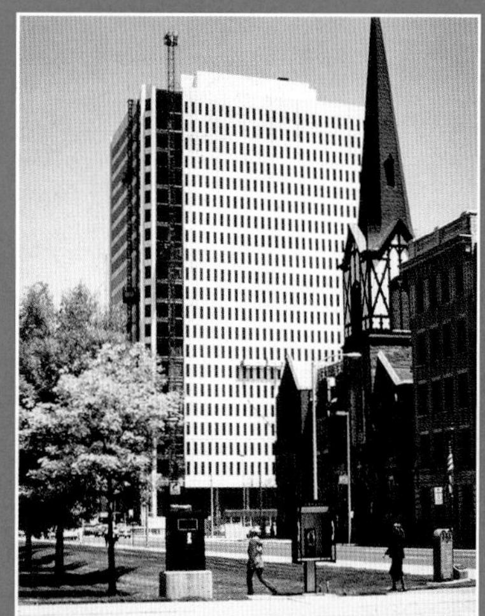

Government Center
Toledo, Ohio
Minoru Yamasaki & Associates
(Minoru Yamasaki)
Once the ideas for the clear span of the perimeter facade and the slightly mitered corners were delivered to Yamasaki, he used them over and over again in other projects.

1981

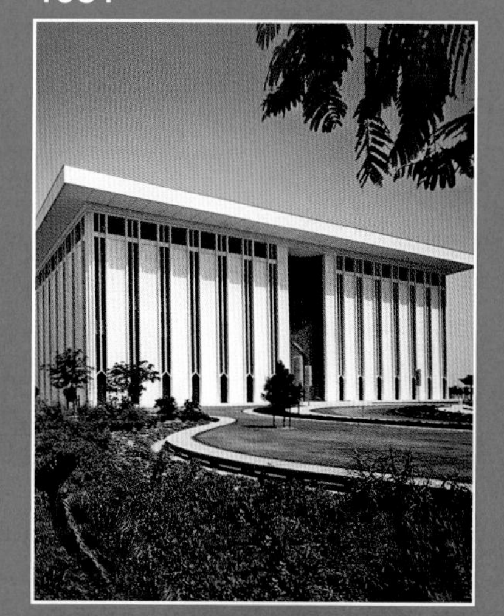

Saudi Arabian Monetary Agency Head Office
Riyadh, Saudi Arabia
Minoru Yamasaki & Associates
(Minoru Yamasaki, William Ku)
The architecture is in some ways emulated by a conservative structural system that attempted to respond to the culture of Saudi Arabia. Our Seattle office carried the brunt of the work, while my participation was largely in communicating with Yamasaki.

1982

Center for Industrial Innovation, Rensselaer Polytechnic Institute
Troy, New York
Mitchell/Giurgola Architects
(Romaldo Giurgola)
The complexity of the facade reflects the complexity of the interior. The structural system takes full advantage of the solid wall panels for lateral resistance to earthquake and wind.

(p.255)

Morton H. Meyerson Symphony Center
Dallas, Texas
I. M. Pei & Partners
(I. M. Pei)

This world-class concert hall was complicated by the need for acoustic isolation of the theater box from a busy nearby airport, mechanical spaces, and the impact of foot traffic. To significantly reduce the size of the curving curtain wall-support girders, thus creating a sense of lightness, the girders are carried by post-tensioning strands buried in the mullions. Among other facets of the design, we carefully studied the theater-related movable systems with the goal of providing a more secure and reliable protection to theater personnel and to theater attendees. The large skylights and sloping glass walls were designed to be structurally redundant.

1982

Terni Fountain Sculpture
Terni, Italy
Beverly Pepper Studio
(Beverly Pepper, sculptor)

Two hollow stainless steel sculptures not structurally connected and separated by 70 feet at their broad but thin bases—create an incredible, cascading waterfall where they seem to meet. The exaggerated cantilevers could not be allowed to touch or to vertically deflect unevenly.

1982

Eastern Province International Airport, West Terminal, Royal Terminal, Mosque, Control Tower
Dhahran, Saudi Arabia
Minoru Yamasaki & Associates
(Minoru Yamasaki)

With a myriad of architectural forms reflecting something of the cultures of Saudi Arabia, the structural systems, many of them exposed, were designed to be equally expressive.

1982

(p.263)

Bank of China Tower
Hong Kong
I. M. Pei & Partners
(I. M. Pei)

We introduced the concept of creating connections for intersecting structural steel space frames within the concrete columns, thus eliminating three-dimensional steelwork. In need of a stalwart structure to resist the severe typhoon winds of Hong Kong, Pei immediately grasped the concept of the perimeter bracing and elected to expose the bracing in the curtain wall design. When completed in 1989, this was the tallest building in Asia and remains today as perhaps one of the region's most elegant.

1983

San Jose Convention Center
San Jose, California
Mitchell/Giurgola Architects, Daniel, Mann, Johnson & Mendenhall
(Romaldo Giurgola)

A clear-span convention center in a high-seismic zone that has already proven its earthquake resistance without difficulty. Robert Shuman of MGA admirably led the production team through the design and construction of this complex building.

1985

South Ferry Plaza Building
New York, New York
Fox & Fowle
(Bruce Fowle)

This ingenious but unbuilt solution for the creation of a ferry terminal and its ancillary facilities was to be constructed beneath and within a proposed garden-like building. Integrated with the structure of the building, the terminal was designed to provide over-water construction access.

1983

Anchorage Historical and Fine Arts Museum
Anchorage, Alaska
Maynard & Partners, Mitchell/Giurgola Architects
(Romaldo Giurgola)

Because of some of the limitations of the construction industry at that time, drawing on technology from the lower 48, this was one of the most technically advanced buildings in Alaska.

1984

AT&T Annex Exhibition Building and Atrium
New York, New York
Johnson/Burgee Architects, Simmons Architects
(Philip Johnson and John Burgee)

We modified Johnson's idea for the curved beams supporting the skylight to be perforated for greater transparency and economy. The massive hanging light fixtures were conceived by Burgee.

1984

Torre Picasso
Madrid, Spain
Minoru Yamasaki & Associates
(Minoru Yamasaki)

This 515-feet-high office building is among the tallest in Madrid. Conventionally designed, the aluminum-clad tower was complicated by the turnover of the entire design and service engineering team, but for us. A large vertical shaft ventilates an underground automobile tunnel.

1985

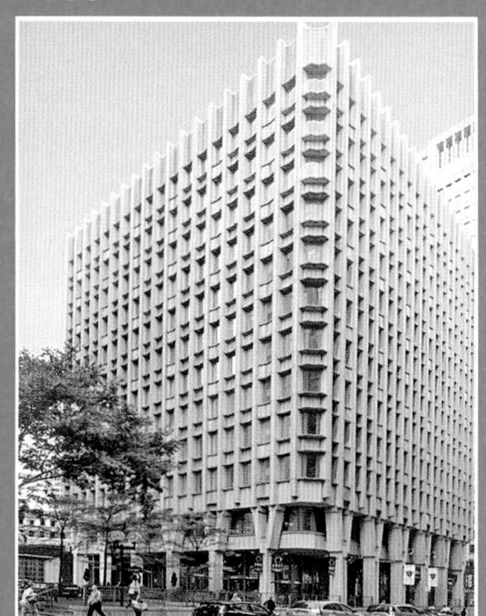

Blue Cross/Blue Shield Building
Boston, Massachusetts
Paul Rudolph
(Paul Rudolph)

Preservation objections to adding floors onto the existing building were mitigated when the developer retained Paul Rudolph, the original designer, as the architect. Originally conceived in an era of very low ceilings, structural and MEP systems were introduced to allow higher ceilings.

1985

Seattle Art Museum
Seattle, Washington
Venturi, Rauch and Scott Brown, Inc., Olson, Sundberg, Architects
(Robert Venturi & Denise Scott Brown)

An elegant design was imposed on this tight and steeply sloped site. Seismic resistance, including isolation systems to protect the works of art, was critical.

1985

(p.321)

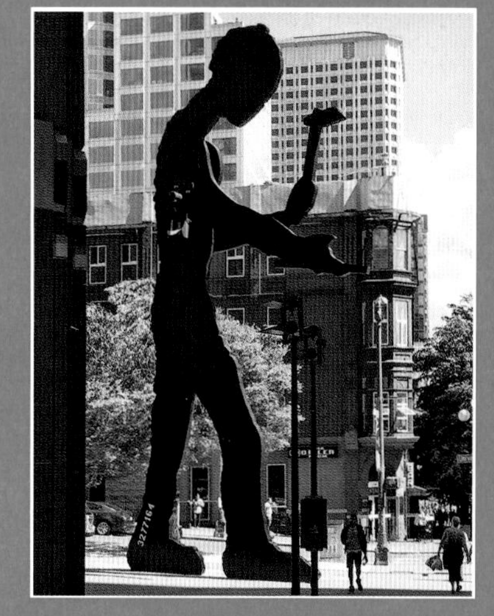

Hammering Man Sculpture
Seattle, Washington
Jonathan Borofsky, sculptor

The first try erecting this kinetic sculpture went awry because the contractor placed the hoisting noose around the neck, causing it to collapse. Following repairs, the noose was repositioned under the armpits and the 48-foot-high hollow steel sculpture was installed without incident.

1986

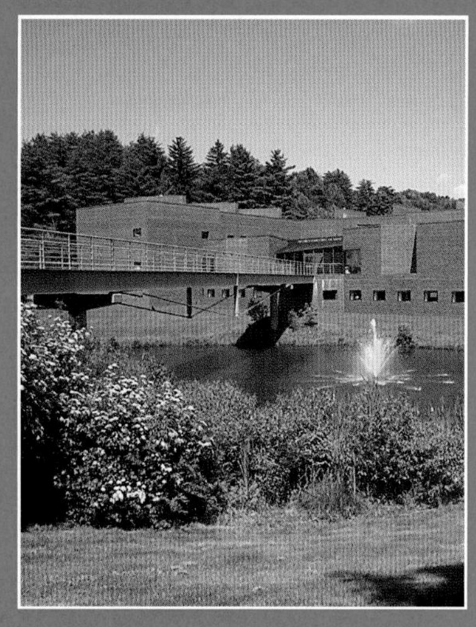

Choate Rosemary Hall Science Center and Bridge
Wallingford, Connecticut
I. M. Pei & Partners
(I. M. Pei)

The engineering challenges were in the brick facade and in the bridge. The facade was specified to be constructed without expansion or control joints while the bridge, spanning a pond, was intended to be elegant within a finite budget.

1986

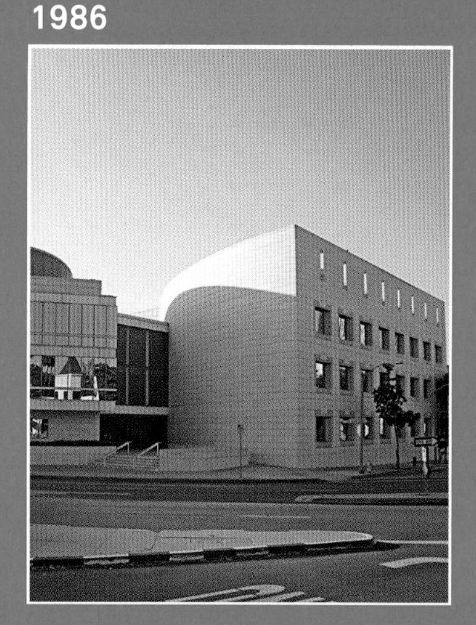

Creative Artists Agency
Beverly Hills, California
I. M. Pei & Partners
(I. M. Pei)

While we produced only the Design Development, thus escaping some of the structural details, the intervention, both extensive and useful, provided constant challenges to the smooth continuation of the work.

1986

(p.241)

Domino's Farms Leaning Tower
Ann Arbor, Michigan
Gunnar Birkerts and Associates
(Gunnar Birkerts)

Birkerts proposed the concept of the leaning building, our first, though never constructed. The challenge was to design a structural system that did not displace laterally under gravity-induced loading. The large model was reportedly constructed without consulting Birkerts.

1987

Grant USA 1
Newark, New Jersey
OWP/P Architects
(Pietro Belluschi)

An unbuilt high-rise, envisioned as the world's tallest, was to be constructed on the relatively weak shale of New Jersey. Without a clear design program, the project seemed to be more of a dream than a reality. While experimenting with the use of diagonal bracing, we produced conventional designs.

1987

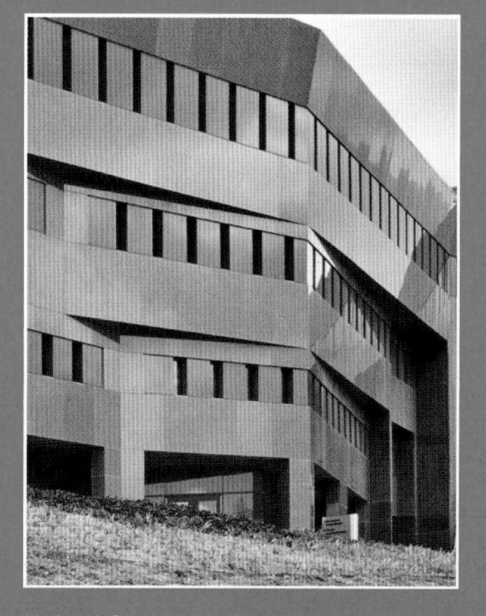

United States Embassy Office Building
Caracas, Venezuela
Gunnar Birkerts and Associates
(Gunnar Birkerts)

The steep hillside site and mandated security provisions and systems, as well as the State Department rules, made the design difficult to achieve.

1988

International Trade Center
Barcelona, Spain
Pei Cobb Freed & Partners
(Henry Cobb)

On a site projecting into Barcelona's Port Vell, this building is in partial conflict with an all-stone berthing pier. Our engineering services extended through Design Development, but we worked diligently to address the issues of the existing pier and those of a tower carrying an overhead tram.

1987

(p.224)

Crystal Cathedral Bell Tower
Anaheim, California
Johnson Burgee Architects, Gin Wong Associates
(Philip Johnson)

The design concept was of a giant upside-down crystal chandelier projecting upward from the ground, not suspended. Johnson intended the "crystals" to be made of glass fitted over a stainless steel structure. But in the seismic zone of Orange County, with a small chapel to be constructed immediately inside and below this system, the risk of falling glass seemed insurmountable. A full-scale polished stainless-steel mockup was produced without glass, convincing Johnson that the new system would be safer and more beautiful than the original design.

1988

(p.275)

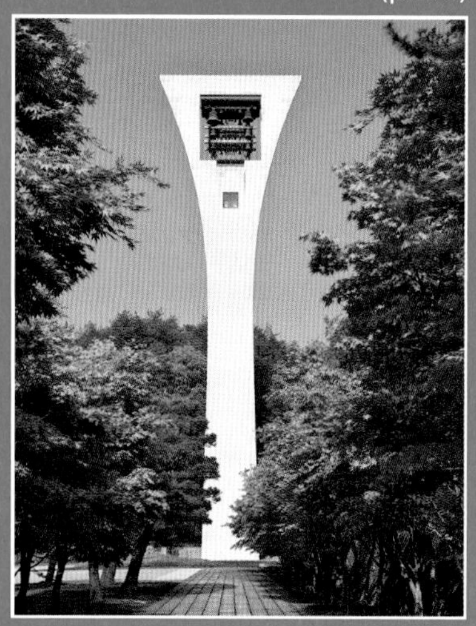

Joy of Angels Bell Tower
Shigaraki Prefecture, Japan
I.M. Pei & Partners
(I. M. Pei)

The majority of the structural engineering being by another firm, we contributed ideas to the fabricator toward the design of the mounts for the bells.

1988

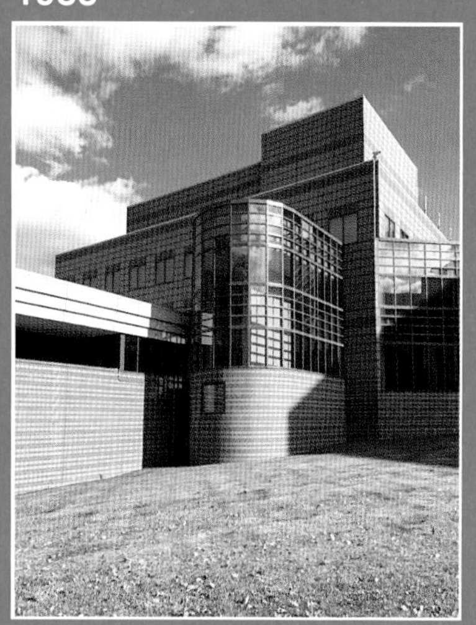

BASF Headquarters
Mount Olive, New Jersey
Grad Associates
(Bernard Grad)

For this campus-style office building, the structure is quite standard overall (structural steel supporting concrete slabs on metal deck), while the long-span structure produced a more interesting architectural design.

1988

(p.241)

Puerta de Europa
Madrid, Spain
Johnson Burgee Architects
(Philip Johnson, John Burgee, and Pedro Sentieri)

For these two leaning, medium-rise office buildings, an ingenious post-tensioning system was developed to deal with the long-term effects of concrete shrinkage and creep, which would have led to further leaning of the building in the long-term. I met with the mayor of Madrid, who expressed a perceptive interest in the design and asked about the base anchorage of the post-tensioning used to "plumb" the building. When the two systems then under study were explained, he directed that the more conservative and expensive system be used: heavy concrete ballasts. While all of the US engineers except us were replaced with Spanish engineers, with the assistance of good language skills in our offices, we could cope with the extensive mechanical/electrical/vertical transportation/plumbing changes initiated by those engineers. A late change by the owner/developer in a raised electrical floor required a reduction in the vertical distance from ceiling to floor. With the HVAC designer unwilling to change his systems, all of that reduction came from the structural systems.

1989

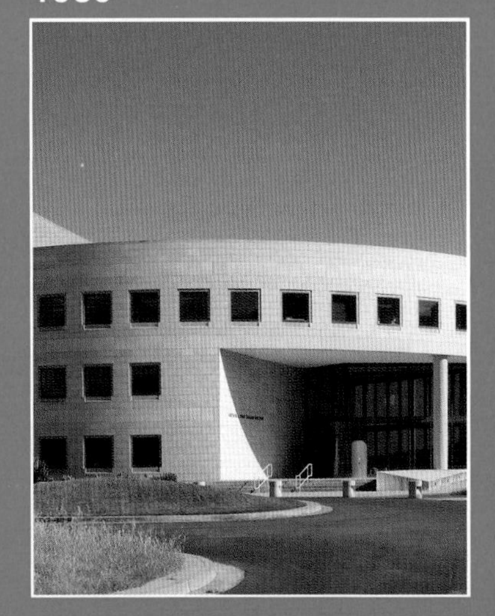

Buck Center for Research in Aging
Novato, California
Pei Cobb Freed & Partners
(I. M. Pei)

The use of hydraulically placed concrete for some of the foundation walls and a seismic gap for the uphill walls perplexed the contractor and the Building Department but both elements proved to be practical and cost-saving.

1990

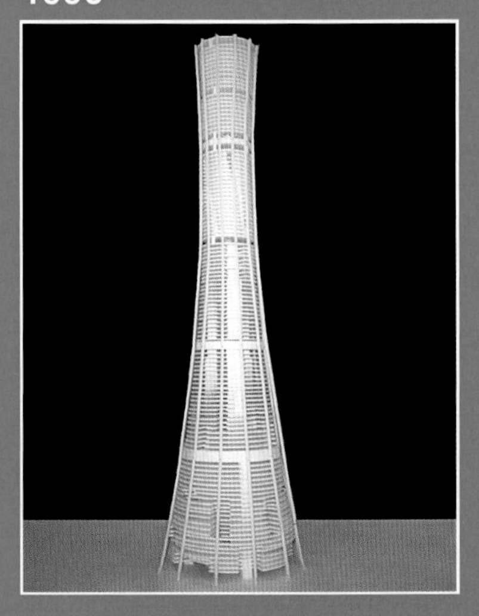

Advanced Building Complex
Tokyo, Japan
ESCO Co.
(Shizuo Harada)

The owner's idea was to include a series of very tall office buildings. We developed the plans, volumetric shapes, and structural systems. For these unbuilt towers, the talented architect, Harada, increased the complexity by adding rapid-transit trains within the proposed building.

1990

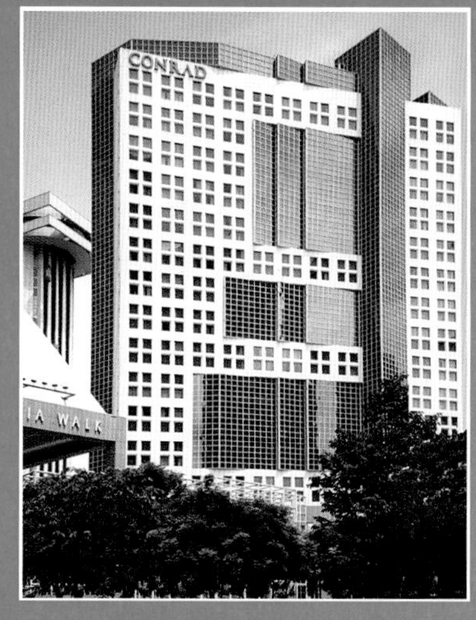

Conrad International Hotel
Pontiac Marina, Singapore
Johnson Burgee Architects
(John Burgee)
Singapore is ideal for tall buildings due to light winds and a low earthquake risk. Burgee's designs for this conventional concrete-frame hotel allowed for a structure that was fast-to-construct and economical.

1990

Millenia Walk Retail Mall
Pontiac Marina Shopping Centre
Singapore
Johnson Burgee Architects
(John Burgee)
While the exposed structure was unusual for Singapore, this upscale 3-story retail mall was imaginatively conceived and planned.

1991

Baltimore Convention
Center Expansion
Baltimore, Maryland
Cochran Stephenson & Donkervoet/
LMN Architects
An innovative three-chord steel truss provides for a 180-foot x 600-foot, column-free exhibition space. This 7,500-square foot expansion tripled the size of the original building, which had been designed by our Jack Christiansen.

1991

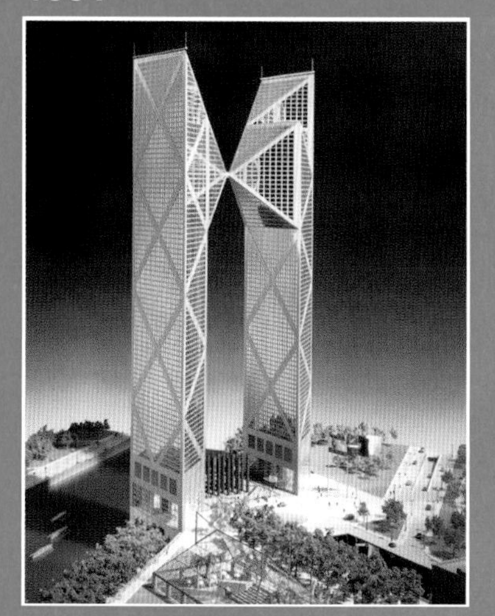

Bilbao Emblematic Buildings
Bilbao, Spain
I. M. Pei Architect
(I. M. Pei)
The building form was entirely by Pei. While we insisted that the "kissing" of the two buildings was not (structurally) appropriate, we completed Schematic Design before the project was halted.

1991

Seattle Art Museum Sculptures
Seattle, Washington
Leslie E. Robertson Associates
(Leslie E. Robertson)
For each of these historic sculptures, we designed base-isolators to mediate the lateral loading imposed by earthquakes, allowing them to move with respect to the building's motion.

1991

(p.276)

Miho Museum
Shigaraki Prefecture, Japan
I. M. Pei Architect
(I. M. Pei)
A tremendous amount of landscape engineering was involved in this project; the top of a mountain ridge was removed to construct the museum, then replaced. In many ways it is a below-grade building. To avoid the mandated use of tile for the roofs, almost all roof areas are skylights.

1991

(p.280)

Miho Museum Bridge
Shigaraki Prefecture, Japan
Leslie E. Robertson Associates
(Leslie E. Robertson with
I. M. Pei, consultant)

This all-steel, post-tensioned bridge features a porous ceramic deck. The pedestrian bridge is strong enough to support light fire trucks or armored cars. The craftsmanship of the Japanese fabrication and erection of the steel works was truly extraordinary, better than anything that we had seen before. Many facets of the siting, design, and construction of this bridge are discussed later in this book, but there are many stories to tell. Very early in the design, our client asked I. M. Pei for a design presentation of the museum, which he was not yet ready to provide. I. M. came to me suggesting that he would have a site model prepared, complete with possible roofs for the museum spaces, as well as the bridge, which was to be the centerpiece of the site model. Almost overnight, we provided the requisite drawings for the model construction. I was astonished to find the exhibition room overflowing with admiring and applauding guests.

1991

(p.255)

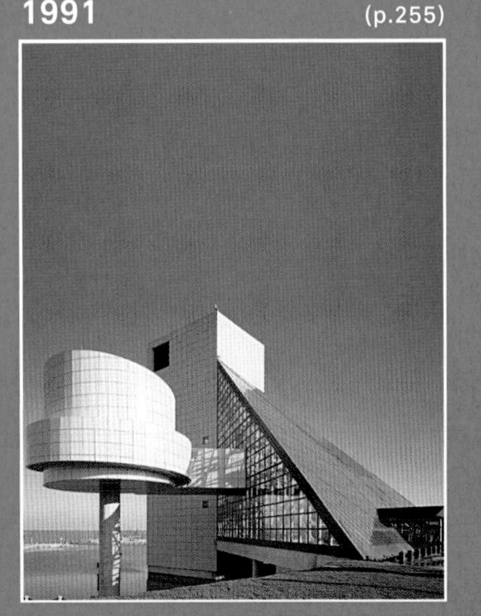

Rock & Roll Hall of Fame + Museum
Cleveland, Ohio
Pei Cobb Freed & Partners
(I. M. Pei)

While the exhibit spaces are below-grade and on land, the Hall of Fame, theater, and related facilities rise directly out of the waters of Lake Erie. To create an almost perfectly planar glass skylight, the large triangulated roof was preloaded prior to erecting the steel, aluminum, and glass.

1992

Reception Pavilion, Miho Museum
Shigaraki Prefecture, Japan
I. M. Pei Architect
(I. M. Pei)

The pavilion is far simpler in both architecture and structure than is the museum, though it enjoys the same fine workmanship.

1994

(p.249)

Saitama Arena Competition
Saitama, Japan
Kikutake Architects
(Kiyonori Kikutake)

This project followed our introduction to Kikutake, who had presided over the induction of the first Honorary Members of the Tokyo Society of Architects, of which I was one. Kikutake's inspiring design for this competition did not win, though on his insistence, we did develop and share a patent for the structural system for the roof of the arena. Our responsibility was for the roofs over the arena and the exhibition space above. When I proposed that our upper roof be turned with respect to the arena below, thus aligning with the city grid, there was considerable skepticism, particularly shown by some on Kikutake's staff. Still, they were good listeners, understanding immediately that this could be done without imposing load onto the arena's roof.

1993

Shun Hing Square, Di Wang
Commercial Complex
Shenzhen, China
Kumagi Gumi
Nippon Steel (client)

In order to reduce the cost and the time of construction, we produced an alternative design for the contractor of the structural steel work.

1993

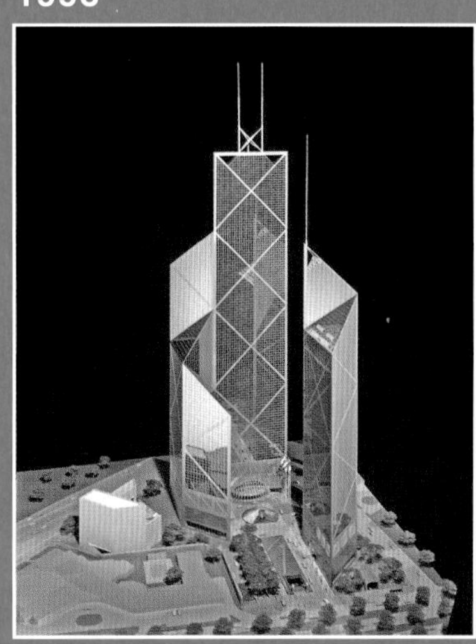

BDNI Center
Jakarta, Indonesia
Pei Partnership Architects
(I. M. Pei and Chien Chung Pei)

Constructed to the 15th floor, this project was stopped by the Asian financial crisis. The structural systems are a part of a family, including the Bank of China Tower. The aesthetic relationship to the Bank of China is apparent.

1994

(p.320)

Manhattan Sentinels
New York, New York
Beverly Pepper Studio
(Beverly Pepper, sculptor)

Located on lower Broadway, two sets of three visually related sculptures surrounded by appropriate trees are sited at opposite ends of a block-long plaza. The work, produced in a foundry not far from Rome, is made of Corten, a weathering steel, thus adding to the corrosion-resistance of the finished product.

1995

Graha Kuningan
Jakarta, Indonesia
Ellerbe Beckett
(Peter Pran)

This 50-story mixed concrete and steel office tower with a 4-story retail podium only made it to the piling before being halted, another victim of the Asian financial crisis. Our difficulties were more in the architectural disagreements than in structural complexities.

1997

International Finance Centre 1 and 2, Retail Bridges, Four Seasons Hotel
Hong Kong
César Pelli & Associates & Rocco Design Architects
(César Pelli)

Located right on Victoria Harbour, this complex has received considerable acclaim and includes the second-tallest building in Hong Kong.

1995

Republic of Korea Permanent Mission to the United Nations
New York, New York
Pei Cobb Freed & Partners
(I. M. Pei)

A small but complex building provides offices, living spaces, and parking for the Korean Mission to the United Nations. Our suggestions for weighing embassy automobiles, as a technique for disclosing the presence of a bomb, fell on deaf ears.

1996

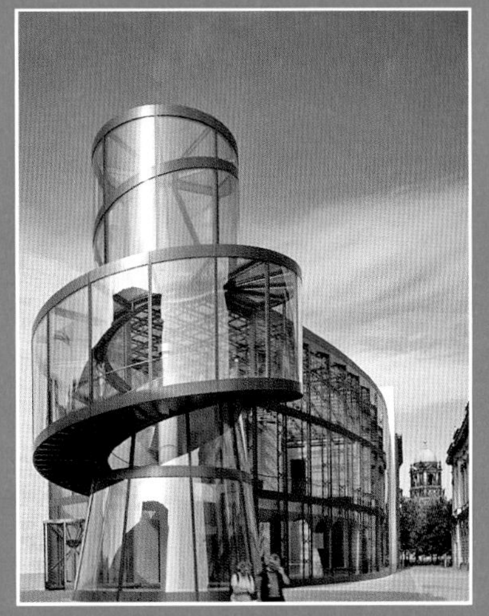

Zeughaus
German Historical Museum
Berlin, Germany
I. M. Pei Architect
(I. M. Pei)

Complete designs were provided for the metal and glasswork, including the twin-glass conical wall of the spiral stair. Concrete work was by German engineers. Concerns about the view through four panes of curving glass proved to be unfounded.

1997

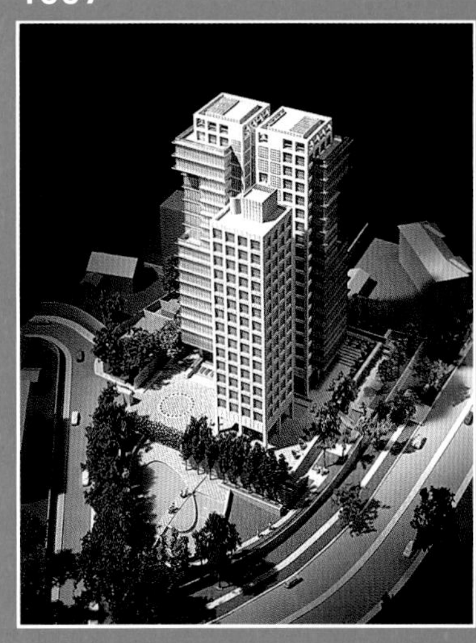

Grange Court Condominiums
Singapore
Pei Partnership Architects
(Li Chung Pei)

This all-concrete residential complex unfortunately remains unbuilt. In part because the wind and earthquake loads of Singapore are modest, the structural system is light. Fallout shelters, required by code, did little to complicate the structural design, but do provide sleeping quarters.

1997

(p.189)

Latvian National Library
Riga, Latvia
Gunnar Birkerts and Associates
(Gunnar Birkerts)

Design development services were provided for what was then the largest building in the Balkans. The design successfully anticipated the rupture of an upstream dam, which could cause flood waters, trees, and debris to impact the facade. Snow removal was a large issue.

1998

Capital Place, Fort Bonifacio
Manila, Philippines
Kohn Pedersen Fox Associates
(John Koga)

An unbuilt, multi-building project on the site of an abandoned US military base, the intended buildings were to display elegant architectural forms with simple structures.

1998

National Constitution Center
Philadelphia, Pennsylvania
Pei Cobb Freed & Partners
(Henry Cobb)

The design of this monumental building contributed to unanticipated complexities in the structural systems, including the huge glass wall that is vertically supported from the roof. In part because of its historical significance, an extra level of attention was devoted to the details.

1998

Espirito Santo Plaza
Miami, Florida
Kohn Pedersen Fox Associates
(William Louie)
A mixed-use building plus parking, Espirito Santo Plaza was designed to withstand hurricane-force winds and a hurricane-induced 15-foot ocean level rise. Temporary struts were designed to bolster the glass and elevator pits were deepened to accept larger pumps.

1998

Cathedral of Hope Bell Tower
Dallas, Texas
Johnson Burgee Architects
(Philip Johnson and
Alan Ritchie)
While hardly a "tower," this was a complex structure with an extraordinarily simple concrete form, conditioned largely by its role as an entrance to the cathedral.

2000

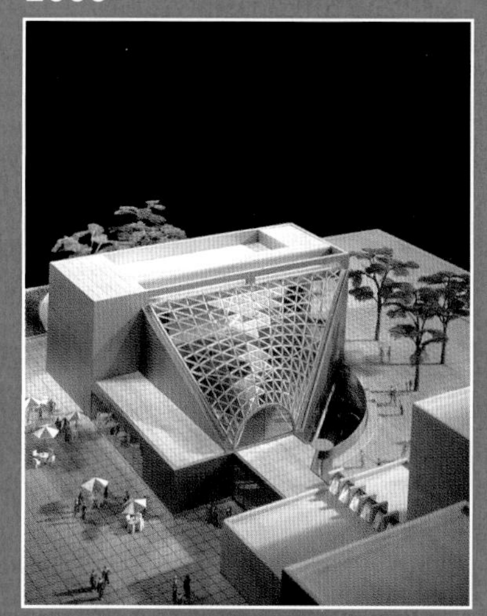

Everson Museum Expansion
Syracuse, New York
Pei Cobb Freed & Partners
(I. M. Pei and John Sullivan)
The skylight was complex, going from flat to arching over an addition to the existing museum. The flat areas of the skylight were thin, requiring exposed post-tensioned cable bracing within the museum galleries.

1999

Oare Pavilion
Wiltshire, England
I. M. Pei Architect
(I. M. Pei)

A 2-story entertainment pavilion on a country residence outside London for Sir Henry Keswick and his wife, Tessa. Pre-deflected cantilevering structural steel, surrounded by concrete, supports the roof.

2000

Bellevue Hospital Ambulatory Care Facility
New York, New York
Pei Cobb Freed & Partners
(Ian Bader)

A new wing with a skylight spans to the existing McKim, Mead & White-designed historic hospital building. For the skylight, the architects accepted the idea of leaning together planar trusses, thus providing the appearance of a space-frame, sans the cost.

2001

(p.301)

Shanghai World Financial Center
Shanghai, China
Kohn Pedersen Fox Associates
(William Pedersen)

Famously, the aperture at the top of the building had been changed from round to that of an isosceles trapezoid. Reportedly, the mayor of Shanghai objected to the round aperture as reflecting the Japanese symbol of the rising sun. The biggest innovation for this 500-meter high, mixed-use building was in providing for a larger and taller building on existing foundations. The weight of the taller building, aggravated by the increase in wind loadings, could not be allowed to increase over the original design. This could only be accomplished by increasing the lateral stiffness of the building.

2000

AIG Tower & Pedestrian Bridge
Hong Kong
Skidmore, Owings & Merrill
(Mustafa Abadan)
This office building features a 45-meter, clear span and curving pedestrian bridge over a broad avenue. The bridge trusses lack an end diagonal and hence support only at one end of diagonally opposite trusses. The trusses do not cantilever; instead they are integrated to form a box truss.

2000

International Commerce Centre
Hong Kong
Kohn Pedersen Fox Associates
(William Pedersen)
A Concept Design was provided for this 108-story, 484-meter tall building on the Hong Kong waterfront, which is currently the tallest on the skyline.

2001

Grand Promenade
Hong Kong
Henderson Land (client)
(WCWP International)
Ideas and Schematic Designs were provided for these five linked, 62-story, all-concrete residential buildings.

2002

(p.291)

Suzhou Museum
Suzhou, China
Pei Partnership Architects
with I. M. Pei
(I. M. Pei. and Li Chung Pei)
With the support of the Local Design Institutes, something of the incredible workmanship found in the Miho Museum was able to be replicated. Well into the project, a major design change swapped all of the tile roofs with the skylights.

2004

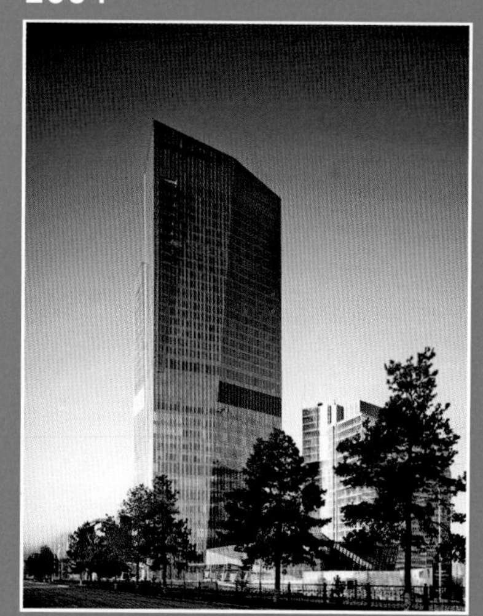

Esentai Tower & JW Marriot
Residences at Esentai Park
Almaty, Kazakhstan
Skidmore, Owings & Merrill
(Mustafa Abadan)
Working in a high-seismic area, coupled with government-regulated, ultraconservative design criteria for seismic resistance, led to ongoing struggles throughout the process.

2005

Kazkommertsbank
Almaty, Kazakhstan
Skidmore, Owings & Merrill
(Mustafa Abadan)
The greatest challenges were the skylight covering the great central atrium and the giant glass wall closing off that atrium. Of interest also were the designs for the bridges crossing the central atrium.

2001

(p.255)

Museum of Islamic Art
Doha, Qatar
I. M. Pei Architect
(I. M. Pei)

This museum seems to float in the waters of Doha Bay, off-shore of the corniche. Instead of building the museum in the water, following our proposal, the contractor first constructed a landfill peninsula of dredged sand. This allowed for a temporary roadway outside of the constructed area, on top of which the museum was to be built. Following construction, the landfill outside of the museum footprint was returned to the seabed. Pei's leadership is clearly visible in the wonderful designs of both architecture and structure. Richard Serra's sculpture 7 was subsequently erected on the peninsula behind.

2005

Esentai Residential Towers
Almaty, Kazakhstan
Skidmore, Owings & Merrill
(Mustafa Abadan)

A set of medium-rise residential towers skillfully designed by the architects and engineered for very high seismic loads.

2005

Beijing Jing Ao Center
Beijing, China
Pelli Clarke Pelli Architects
Henderson Land (client)

Idea-generation and Schematic Design were provided for this faceted building complex. In the end the structural systems, while far from conventional, were well within the technology.

2006

Mabarak Center
Lahore, Pakistan
HOK
(William Hellmuth)
Before the project was halted, only piling had been constructed for this complex of earthquake-resistant office and residential towers. Beyond major high-rise buildings, a giant convention hall and a related power plant, we advised on emergency water storage and similar safety features.

2006

Rotating Tower
Dubai, UAE
David Fisher
(David Fisher)
Different from that of a traditional rotating restaurant, Fisher's concept was to have each floor rotatable with respect to the floors above and below, requiring that each floor be cantilevered on rails from a central services core. While still unbuilt, further study is ongoing.

2007

Nakheel Tall Tower
Dubai, UAE
Woods Bagot
(Woods Bagot)
Continuing from one of our prior high-rise projects in Dubai, this proposed but partially constructed (piling only) kilometer-high mixed-use tower remains idle. At the architect's insistence, the use of three vertical slots through the building was changed to four slots, with an accompanying loss of effectiveness in resisting wind loads. Our contribution to the continued project is limited to that of an advisor/consultant.

2007

Mary Avenue Pedestrian/Bicycle Bridge
Cupertino, California
HNTB Corporation
City of Cupertino (client)

After the project had been designed and tendered, cost-savings measures were requested. We proposed a variety of changes, the most important being that the construction be changed from reinforced concrete to structural steel, reducing the cost by half. Our peer review and value engineering services were carried through Design Development, with the structural engineering completed by others.

2007

Shenyang International
Financial Center
Shenyang, China
Pei Partnership Architects
(Li Chung Pei)

This would have been the tallest building in Shenyang. Ground was broken but the city, ultimately, did not authorize construction. Concept Design services were provided, in sufficient detail to allow an evaluation by the government.

2007

Ritz-Carlton Astana
Astana, Kazakhstan
Robert A.M. Stern Architects
(Megan McDermott, and
Robert A.M. Stern)

An unbuilt hotel features a complex skylight. Bob Stern sat in on almost every one of our project meetings, each time bringing his focus on areas deserving of more in-depth thought. The need for seismic resistance was paramount.

2007

(p.286)

Miho Institute of Aesthetics Chapel
Shigaraki Prefecture, Japan
I. M. Pei Architect
(I. M. Pei)

Services through Schematic Design were provided for this beautiful chapel. This was I. M. Pei's last project before his retirement. I was able to (softly) convince I. M. that his design for a pleated roof should be changed to that of a smooth, curving surface, thus simplifying the construction and reducing costs. Completed architectural and structural designs were accomplished by a joint US/Japan group.

2009

Lotte World Tower
Seoul, South Korea
Kohn Pedersen Fox Associates
(James von Klemperer)

At 555 meters, this is currently the tallest building in Korea. Construction on the mixed steel-and-concrete tower topped out in 2016. As of this writing, further details are not permitted to be disclosed.

2009

United States Embassy
London, England
Pei Cobb Freed & Partners
(Henry Cobb)

This entry did not win the competition, but the designs were truly wonderful and Cobb's concept exceptional. Of course, the State Department demanded and received appropriate levels of security, redundancy, and robustness.

2009

NASCAR Hall of Fame and Museum
Charlotte, North Carolina
Pei Cobb Freed & Partners
(Yvonne Szeto)

With its beautifully curved roof/facade and the inclining racecar display surfaces that mimic the sinuousness of a racetrack, the museum's exhibition space was challenging even to our highly expert design team. Beyond the tortuous structure for the exhibition space, the other buildings are comparatively simple.

2009

(p. 318)

7 Sculpture
Doha, Qatar
Richard Serra, sculptor
(Richard Serra)

Beautiful and gutsy, this sculpture consists of seven 200-millimeter thick steel slabs, each 24-meters high by 2.4-meters wide, connected only at the foundations and at the top. The slabs are carried on sliding supports as required by the severe temperature differentials augmented by solar radiation.

2010

Busan Lotte Tower
Busan, Korea
Skidmore, Owings & Merrill
(Lotte Group)

Rising 510 meters, we provided peer reviews, value engineering, and alternative designs.

2010

CTF Tianjin
Tianjin, China
Skidmore, Owings & Merrill
and ECADI
(Tianjin New World Huan Bo Hal
Real Estate)
Peer review, value engineering, and alternative designs were provided for this 530-meter high mixed-use building.

2010

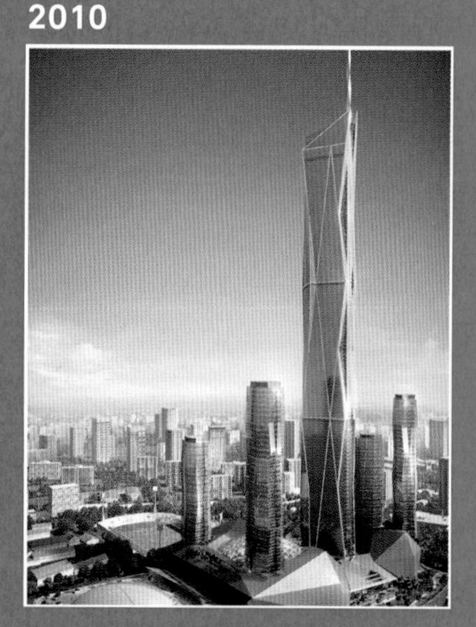

Merdeka PNB118
Kuala Lumpur, Malaysia
Fender Katsalidis Architect
(Karl Fender)
When completed, this articulated mixed-use tower, at more than 630 meters, will be the tallest building in Malaysia. Along with Robert Gird Group, we provided a strong Design Development followed by a peer review of the local engineer's construction documents.

2011

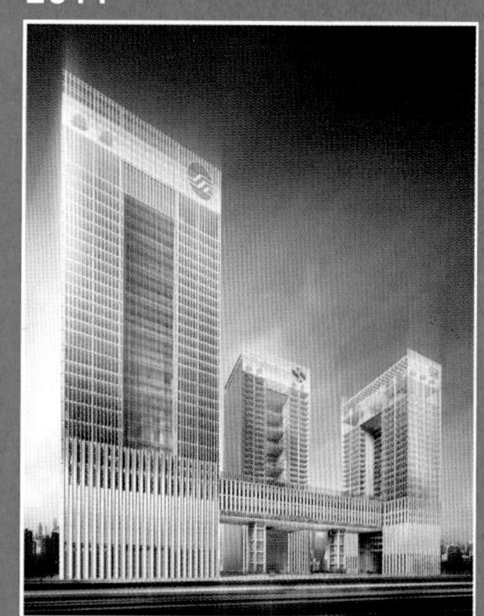

Shanghai International
Financial Center
Shanghai, China
Murphy/Jahn Architects
(Helmut Jahn)
Peer reviews, value engineering, and an alternative design were provided. Because of differences between the designated structural engineers and the expert panel, the proposed alternative design was carried forward.

2011

Suzhou Pedestrian Bridges
Suzhou, China
SWA Landscape Architect, Leslie E.
Robertson Associates
(SawTeen See and Leslie Robertson)
Two park-like bridges connect large commercial terraces to parks located across the roadway. These are post-tensioned, exposed-concrete bridges with randomly located large openings to the roadway below. The root-balls for the trees are buried within the structure.

2012

Shenyang Office Building
Shenyang, China
Aedas
(Keith Griffiths)
This was a competition for a 520-meter, mixed-use building, all of structural concrete except for the floor framing outside of the services core. The structure also included outrigger and belt trusses as well as perimeter moment frames and bracing.

2015

470 11th Avenue
New York, New York
Kohn Pedersen Fox Associates
(David Iwami Malott)
In competition for a high-rise in New York, this unusual project featured exposed bracing incorporated into an imaginative architectural design. While not selected for construction, both the architectural and the structural designs were exceptional.

(p.311)

NK Sky Tower
Tokyo, Japan
NHK TV/Kohn Pedersen Fox Associates
(David Iwami Malott)

This was a study of a theoretically possible super high-rise to be built at some time in the future. The concept was part of an initiative called "Next Tokyo" and was broadcast on Japanese television, quickly attracting worldwide attention. The proposed structural designs, unlike anything constructed in the past, include sets of three buildings rotated 60° with respect to the three buildings both above and below. The system is uniquely open, allowing the free flow of wind-driven air between the sets of three buildings, creating a high level of aerodynamic damping and thus reducing the wind-created oscillation of the towers.

II

Life and Philosophy

EARLY YEARS

INTRODUCTION

Being a terrible student, though always stellar in conceptualizing systems and shapes until I was well into university, I had no understanding that there was such a thing as a "structural engineer." The rocky road to that profession meandered through my service in the US Navy, a disjointed passage through the University of California at Berkeley, intermittent work as a school teacher, stints racing cars, coauthoring unpublished children's books, apprenticing as an electrical engineer, and more. While all of my searching for a niche in life may seem to be random, it was a joyous path that I followed. Once settling on the discipline of structural engineering, I have learned to warp that path to include contributions into architecture and pacifist and human rights activism. It has been a most fulfilling career.

In recounting the roads that I have traversed, we should begin with the experiences of my youth.

I was born in Los Angeles in 1928 to Tinibel Flora Grantham and Garnett Roy Robertson; my brother Taylor (Ted) was a year-and-a-half older. My mother was beautiful, and as I understand it, an orphan, while my father came from a family of successful businessmen. He was a jack-of-all-trades who had a host of jobs: acoustician, carpenter, house painter, inventor, machinist, manager, rancher, salesman, seaman, wine engineer, and more. With the nickname "Hap," he was an upbeat kind of guy. His father, LeRoy Robertson, was a rancher and banker. In his later years he was a bank manager. He and his wife, Ethel E. Spears Robertson, were financially strong,

Below Grandfather LeRoy Robertson (standing left) with other ranchers in Montana c. 1890.

Below right A rare family photo with father Garnett "Hap" Robertson and birth-mother Tinibel Grantham Robertson; Leslie on the left and brother Taylor (Ted) on the right.

making it possible for my father, as a boy and as a young man, to live comfortably. With the death of my grandfather in 1921 and the onset of the Great Depression years later, all was to change.

My earliest recollections, from about age four, are of our living in Manhattan Beach, California. Mother, Ted and I lived with Grandmother Robertson in a building consisting of two apartments atop a parking floor, separated from the beach by a street and a railroad track. Grandmother occupied one apartment, while we occupied the other. Around this time I believe that my parents had separated. It's hard to remember my mother. While I know virtually nothing of the divorce, it seems that, before or after, she became infatuated with another man, a director of cartoon animation at Disney Studios whom she later married; I never met him.

The apartment we continued to live in was owned by a relative, William Taylor (Tay) Garnett, a successful director and writer of films (including *Skyscraper* in the year of my birth as well as major films), and of TV. While our room didn't have a radio, every night Ted and I listened to *The Lone Ranger* from our grandmother's apartment, cheering on his exploits.

Later, we moved to downtown Los Angeles, there to reside in a multistory hotel, which our grandmother managed. In that environment there were no other children, drawing Ted and me closer together. Grandmother, who was a brilliant and loving woman, intermittently raised Ted and me throughout our childhoods, but at some point we were sent to a live-in military academy, where discipline was stressed. I was put back one grade so that Ted and I, in separate quarters, hardly saw each other. It was my first experience with racial bigots, with homosexuals, and with the love of power and money. I deeply hated that school, leaving at the end of the school year but without the intended inculcation of "discipline."

We next went to live with our mother elsewhere in downtown Los Angeles, with rooms in the back of a restaurant. I remember getting lost after my first day of school but found my way home by methodically following the city grid.

For reasons unknown to me, despite my poor academic performances, I was advanced two grades, putting me just one grade behind my brother.

Somewhere in this time frame we met our Aunt Rade, the mother of Tay Garnett, who must have provided us with additional financial support. She owned a large black limousine. I did enjoy my rich relatives! But on one occasion, we stopped for lunch, there discovering, to my amazement and concern, that my friend the chauffeur, being black, was not permitted to dine with us.

I was seven or eight when our mother sent us north by train to the San Francisco Bay Area to spend the summer with our father. By that time Dad had married a wonderful woman, who loved and served him until his death at age ninety.

Left Older and taller brother Taylor and me in uniform during our one-year stay at a military academy c. 1935.

Right As boy scouts with Grandmother Ethel Robertson in Lafayette, California.

We lived in a tiny cabin, in a redwood canyon, up a long path from the valley floor. With exploring the nearby hills and romping with my stepmother's dog, Blanco, it was a great summer. Years later, bicycling through that same redwood valley, the man who owned the area told me that we had skipped out without paying the rent. As we were very poor, with little luggage to carry, such is not unlikely.

At the end of summer, Ted and I rode the train back to our mother and to the rooms behind the restaurant. As usual, I did poorly in school, perhaps related to my ability to assimilate concepts but my failures in rote learning.

Nearing the end of the school year, our mother explained that she was not able to deal with us and that we were to return permanently to the care of our father. It's not unlikely that the "dealing with us" was really dealing with me. And so it was that we again rode the train north to the San Francisco Bay Area. This was the last time that I saw my mother.

Our father and stepmother were then living with others in a big old house on a large piece of property near downtown Oakland. It was my first extended experience with non-Caucasian (i.e., black) people. For reasons of his own, Dad bused us to an all-white public school. Living in this old house, it was difficult to meet or relate to the other children in our neighborhood.

The most important part of this era is that my father's wife had evolved into becoming my mother. For the rest of her life and until this very day, I think of her as Mom. Of interest, Ted always used her given name, Zelda. I understand that Ted later remained in close touch with our natural mother and with her husband; perhaps, because she seemed more like a nurse than a mother and because of the contributions of Grandmother into my life, I had no interest in contacting her.

The Great Depression continued until the onset of World War II. Jobs were nigh impossible to find. We did not go to the soup kitchens nor did we go hungry—this again was perhaps with Grandmother's help.

At age nine or ten, I got my first paying job, helping a blind woman who lived two blocks away. We called her Aunt Eunice, and after school I would walk her dog, assist in cleaning her house, shop, and take her on short walks. She had a large 78 rpm phonograph, all mechanical with fiber needles. It was my first real introduction to music of any kind but, particularly, Mozart, Brahms, Beethoven, Haydn, Mozart, Schubert, and others. Oh, how I loved to listen to that music! It was for me an overwhelming experience, lovingly tolerated by Aunt Eunice.

With Dad and "Mom" (stepmother Zelda) in the mid-1930s.

By the time I reached the 8th grade and Ted went on to high school, we had moved again, this time to another Bay Area town, Lafayette. After a year, because I was such a poor student, and young for that level of schooling, my teachers wanted me to repeat 8th grade, but Dad intervened and I went on to high school. After school and on weekends, much to my satisfaction, I started working in a gasoline station, pumping gas, washing windshields, checking and changing oil, washing and lubricating cars, and otherwise making myself useful. It was my kind of work, and exhilarating to be able to contribute to the family income.

For whatever reason, perhaps because my brother was such a good student, the administrators put me into classes with boys and girls who were both older and better students. For example, I suffered through a year of Spanish, but everyone else in the class had taken Latin and they were a year or two ahead of me in school. I hated the class, achieving the lowest grade in each and every examination.

In my freshman year, barely a teenager, I quit my gasoline station job to become manager of a competing station. I was proud of my duties: opening the station, closing up in the evening, putting the accumulated moneys and the gasoline rationing stamps into a little safe, balancing the books, cleaning up, and the like. It's important to remember that, with World War II, there were not a lot of men around who could do the kinds of things that cars needed to have done.

At school, I dug in my heels to major in that which I now call "shop and girls"; shop was the part dealing with the learning of woodwork, mechanical drawing, and auto repair, while girls was the fun part, though I was far too homely and clumsy to do very well. I recall that, in mechanical drawings, being gifted with a talent for visualizing objects, I completed more than twice as many drawings as others in the same class, which our teacher recognized by giving me two credits instead of one. Hey, while I was close-to-failing in all of my other classes, I received an "A" twice.

Because of World War II, I volunteered after work as a sort of air-raid warden. My job was to watch and to report on the path of all aircraft passing over our town. I was given a little book, *IDENTIFICATION OF AIRCRAFT FOR ARMY AIR FORCES GROUND OBSERVER CORPS*, which I handled with loving care. I was proud in believing that I was defending our country.

On completing high school, Ted enrolled in a US Navy officer-candidate school to become an airman. After limping through three of the four years of high school, I dropped out and having just turned sixteen, lied about my age to join the Navy as a seaman.

THE NAVY

My first base was the Great Lakes Naval Training Station, north of Chicago, where I was selected to go to electronic technician's mate school, a one-month weeding-out group designed to select those best suited to learn the repair of radar, radio, and sonar. Then, it was on to Del Monte, California where, pushed by my Commanding Officer, I took the examinations for Annapolis. Much to my astonishment I was selected! This netted me a discharge from the Navy along with two weeks leave, complete with instructions and railroad tickets to report for duty at Annapolis, Maryland.

I had no interest in becoming a Naval officer and had no understanding of the nature of Annapolis. But I made full use of my two weeks leave by spending it in Hollywood with Tay Garnett, who was then directing *The Postman Always Rings Twice* with Lana Turner. He even arranged for me a dinner date with Shirley Temple; we were of the same age, but she was a sophisticated young lady while I was a country hick; there was absolutely nothing in common between us. She was kind and ladylike enough to stay through dinner.

I returned to my Del Monte base, cap in hand, to explain to my Commandant that Annapolis was not for me. Fortunately he, being something of the fatherly type, re-enlisted me into the Navy. I did pretty well, rising to Petty Officer, Electronics

Technicians Mate 3rd Class. Mine was a calm Navy experience in that I never heard a shot fired in anger. Still, from the effects of a modest explosion, my left knee was damaged; I lived with it for sixty years before having a full knee replacement.

During this period of World War II, I was profoundly affected by the deaths of three young men (more realistically, they were boys). One was killed by a kamikaze pilot in the Pacific; one failed to make the top of the cliffs at Normandy Beach; while the third, who was Japanese-American and born in the United States, was sent to an internment camp in Idaho, only to be killed in a fall from his tractor. The senseless loss of those I had known and loved drove me resolutely down a path of what, in time, I came to know as pacificism.

Electronic Technician's Mate relaxing while on shore leave.

THE UNIVERSITY EXPERIENCE

While my interests had always been in shop, racing cars, mechanics, and the like, upon being discharged I learned that I could attend a university at the expense of the government under the G.I. Bill. I knew nothing about universities and had no interest or even understanding of structural engineering, but with so many young men coming back into the work place, there were few jobs, making additional education an attractive alternative. I studied hard, learning things that I had failed to learn in high school. It was amazing, and stimulating to me that a country boy without a high school diploma could gain entrance into the prestigious University of California, Berkeley.

I had other experiences that informed my professional life, even though they weren't jobs per se. Early on, I collaborated with a talented artist, a young woman, on children's books. I would make up a story, she would block out the illustrations, we would modify the story to fit the illustrations and so forth. We submitted to publishers but none were accepted. We learned right away that children's books are written for the adults who purchase them, not for young readers, and that I was no Hemingway and she no Rembrandt. Times were changing; Jack Kerouac and Jasper Johns were carrying the day. We were good friends but reluctantly, she and I went our separate ways. I had more success off the books, as it were. In the Navy, playing cards, I had earned a fair amount of money; I wouldn't call it gambling because I just outsmarted the other sailors in counting cards and that sort of thing. Putting together my winnings, plus money that I'd earned from military service, funds from the G.I. Bill, waiting on tables and doing dishes in a sorority, and summer jobs, I had plenty of money and no student loans.

On the weekends I earned a little money driving racing cars at county fairs. I did better in demolition derbies, where survival of one's vehicle was overshadowed only by the need to put on a good show for the audience. In truth, on the financial side, beyond that paid to the owners of the prize-winning cars, we split the dollar prize equally among the drivers, regardless of our standing at the finish line. We raced hard, but without the risks that might have been generated in seeking the winner's purse.

Biased by my experiences in the Navy, I thought that I would major in some combination of mathematics and electrical engineering. Even so, at Berkeley, it was easy to partake of classes in other colleges. Once, the famous architect Frank Lloyd Wright came to lecture. He was an imposing figure who, having created a host of incredibly beautiful designs, even to this inexperienced student, could be seen as an extraordinary talent. Professors aside, Mr. Wright was the first real architect that I'd met. Enlightened by military service, I was turned off by his self-centeredness, his almost lecherous approach to the women attending the lectures, and by my perception of his attitudes toward the suffering world.

Frank Lloyd Wright with students at the School of Architecture, UC Berkeley, April 1949.

For inexplicable reasons, Wright somehow sensed that I was unlike many of the other students, that my mind was older than my body, and with no way of knowing that I was not enrolled in architecture, immediately prior to his departure took me aside to say that I would learn nothing at Berkeley, that I should work for him at Taliesin West. Taking a deep breath, and braving his wrath, I told him I had learned from his personal example that architecture, as a profession, was not for me—a decision I have never regretted.

EARLY PROJECTS 1952–1964

KAISER ENGINEERS

On graduating from UC Berkeley in 1952, with no time for the commencement exercises, job hunting was heavy on my mind. Hunt I may, but all of the jobs that I could find were in the so-called defense industries which were really offense industries; I was not going down that road. I finally landed a job as a mathematician in the Electrical Department of a contracting company, Kaiser Engineers. Commuting by bicycle from Berkeley to Oakland, my starting salary was well under $70 per week, but salary increases came quickly.

My first task was to sort out how to economically distribute electrical power in large, complex electrical grids. It seemed that I had the mathematical tools not enjoyed by the electrical engineers. In any event my efforts proved to be a smashing hit with the Electrical Department, giving me considerable leeway as to that which was to occupy my time. My next task, then, was to sort out something of the suspended transmission lines distributing that power, the most interesting problem being the circumstance where a wire breaks or an insulator fails. What is the dynamic impact on the remaining wires and on the towers? Not having computers, it was possible to muster a group of women with desk calculators to sort out solutions to these problems. While these were not precise solutions, they were good solutions, far better than the guesstimated solutions that the electrical engineers had been working with.

It was then that I became interested in the effects of these loads from the broken transmission wires on the design of the transmission towers. The Chief Engineer, Bill Bertwell, who was a structural engineer, listened kindly to my story, voicing the opinion that, just perhaps, I could do useful things in the Structural Department. Accordingly, I set out to develop design parameters for the steel towers. This required the accumulation of knowledge regarding the design of structural systems and of individual components. This expanded into a series of other structural problems. In order to accomplish these tasks, it was necessary to work very hard to learn about structural engineering; Bill Bertwell was most encouraging and more than a little helpful.

While I studied in the quiet of my evenings and weekends, the computational tools of that time—slope deflection, Hardy Cross analyses, and the like—Bill would lay aside more important work so as to guide me through those systems. Of even more importance, he introduced me to the graphical methods of analyses of structural systems. These graphical solutions were, to me, captivating in ways beyond my imagination. I could, with a parallel rule, a few plastic triangles, and an engineer's three-sided scale, analyze structures that seemed otherwise impossible to this fledgling structural engineer.

Eventually, a call came from Bill Bertwell who explained that the Structural Department was going to be permanently closed, adding that I was to be released; very generously, he introduced me to the firm of John A. Blume & Associates.

JOHN A. BLUME & ASSOCIATES

While I still don't understand the corporate structure, John A. Blume & Associates was made up of a series of individual corporations. I ended up being the sole employee of the Research Corporation and one of the many employees in the Design Corporation, with half time in each. As well, he had a Drafting Corporation, an Engineering Corporation, an Administration Corporation, perhaps others. I worked on a museum to be located on the most southerly tip of the North Island of New Zealand. For this project I was a minor developer for the designs for earthquake resistance and for a 6-meter rise in the height of the sea. It was, to me, fascinating work.

I also worked on some offshore structures, perhaps the most interesting being a structure by the name Rincon Island, to be constructed off the southern coast of California. Being a landfill, to fend off the impact of the ocean waves, we surfaced the edge of the island with tetrapods, these being huge concrete members with four equal arms such that they were all in different planes but with a common intersection at the center of the tetrapod. Each of the arms, conical in nature, circular in cross-section, were small at their tips but larger as they intersected the central sphere. I can remember sorting out how to calculate the precise amount of concrete in one of those tetrapods. The solution came to me on the day before Christmas. I remember screaming, bringing people from all over the office to sort out the commotion. Beyond such trivia, I worked up both experimental and testing methodologies for the stability of the tetrapods against the on-coming ocean waves.

A bridge was required to reach the island, with the design falling on my shoulders. Since it was a narrow, one-lane bridge, there was no passing of vehicles. This led to my idea that such eccentricities of load from the center of the piling could be handled easily with bending in that piling, while the lateral forces from earthquake and ocean waves required battered piles. My idea was to alternate bays of single vertical piles with bays containing two battered piles; the latter to resist the lateral loads from earthquakes, wind, and waves. Very generously, John Blume allowed my designs to be constructed. While the bridge may be in use to this day, during construction and in a heavy storm, witnesses attest that the pile driver, supported on a cantilever from the permanent piling, swayed in the wind before disappearing into the sea below. Fortunately, such matters were left to the contractor.

Above A solitary figure provides human scale for the huge four-armed concrete tetrapods installed as breakwaters against incoming waves.

Top right The 2,700-foot-long bridge to Rincon Island, California, rests on single vertical piles with alternating pairs of battered piles to resist wind, waves, and earthquakes.

I soon came to realize that my salary was substantially lower than others who were working under my direction. Accordingly I went to see John Blume, telling him that I felt that a salary increase was only fair. He said "Well, how much of a raise do you believe to be appropriate?" In truth, I really hadn't thought about it. I blurted out that a doubling of my salary was appropriate. Several days passed when, quite by accident, I overheard John and his three lead engineers discussing the matter. John was sort of wishy-washy, not being sure that I deserved the full amount, while the threesome was quite negative. On the following day, John called me to his office to say that they had talked it over and agreed on a 50% increase. Knowing that this was coming, I had thought about it carefully, deciding that I deserved the 100% increase. Accordingly, I told him that I had inadvertently overheard his conversation, and that Raymond International had offered to triple my salary plus all moving expenses to their headquarters in New York. Sadly, as John was an exceptional engineer, one that I admired and respected, we parted.

RAYMOND INTERNATIONAL

Raymond International was basically a pile fabricating/driving organization. They had constructed a series of offshore platforms in Lake Maracaibo, Venezuela, one of which had collapsed. The sole survivor said that the platform shook, there was a sound similar to the breaking of wood, and he woke up in the water. This book is not the place to explain such structural problems but, happily, I was able to sort out the causative factors. I also worked on their alternative design for the foundations of the General Rafael Urdaneta Bridge, in Maracaibo, which was a very heavy concrete suspension structure. Fortunately, I was able to introduce a series of ideas on how to make these massive foundations more economical while, at the same time, faster to construct.

As the Chief Engineer seemed to believe that I was striking for his job—inconceivable to this not-yet-thirty-year-old engineer, complicated by his marriage to the daughter of the Chairman of the Board, with whom I had developed a father/son kind of relationship—this was not a happy time for me. Following the fulfillment of my one-year contract, I piled a tent and sleeping bag into a red VW bug convertible and headed west. I ran out of gasoline and money in Seattle, where I was able to obtain an advance on my salary from a new job.

My employer in Seattle was a firm called Worthington and Skilling but, in fact, I was the sole employee in a next-door office under the direction of George Runciman. While George and John Skilling were an incompatible pair, George and I were well-matched, he being a completely practical person while I leaned more to the theoretical side. Perhaps of interest, unlike other engineers who used a note pad for structural calculations, George made his calculations right on the drawings. His cousin made all of the drawings while I developed the structural details and checked George's calculations. I learned so much from George, both in engineering and in his wonderful manner of dealing with men and women of all ages and backgrounds.

George, reeking of practicality and honesty, was completely believable as an expert witness in court. You could almost see the hayseed straw projecting from his mouth. While his theoretical background left something to be desired, he had a sound understanding of structures.

Perhaps because of the differences in our respective talents, he introduced me to his clients as: "This is Les Robertson, my California screwball." While at first

The base of one of the piers of the General Rafael Urdaneta Bridge in Venezuela, facilitated by this engineer's ideas on how to make the massive foundations more economical and faster to construct.

In the Seattle offices of Skilling Helle Christiansen Robertson with firm leadership: Jack Christiansen (left), John Skilling (center, pointing), and Les Robertson (right).

taken aback, I immediately learned that this was quite a compliment coming from the George that I had come to admire and to love.

As the months passed, George decided on retirement, sadly, leaving me to cross the hall into the offices of Worthington and Skilling. This led to my change in focus from industrial bridges and buildings to that of buildings founded on architectural concepts, raising my sights into a whole new world of design.

SEATTLE CONVENTION CENTER AND OPERA HOUSE, 1960–62

One of my more interesting projects was the turning of an existing big concrete box of a building into a concert/convention center, later known as the Seattle Convention Center and Opera House. B. Marcus "Benny" Priteca, who had learned the art of theater design in vaudeville houses on the West Coast was the lead architect. While he was much older and culturally we were far apart, we became fast friends.

One day Benny announced that we needed an elevator from the stage to the fly gallery. I was quick to point out that we didn't have the money for that elevator. Still, Benny insisted. Seeking to support my negative position, pointing to a shear concrete wall, with embedded u-shaped iron rungs up to the fly gallery, I said "Okay, if I climb that ladder, will you agree that the elevator is not needed, that the stage hands are able to use the stairs?" Benny agreed. While I climbed that very tall ladder, with no fall protection or safety cage, it should be admitted that my heart was pounding. The vice-like grip of my fingers may have left dents in the rusty iron rungs of that ladder.

Many years later, by then I was in New York for the World Trade Center, Benny telephoned to say that there was a problem with the stage. Returning to his days working with vaudeville theaters, he had designed the wooden stage floor for a high-dive act, or for the passage of an elephant. The floor was strong. The floor was stiff. The floor was everything that ballet dancers did not want. In short, dance companies refused to perform on Benny's stage. His call to me was in desperation.

As luck would have it, I had dined a few times with the late Violette Verdy (Nelly Armande Guillerm), prima ballerina of the New York City Ballet, whom I had met on a flight from someplace to New York. Over dinner, we discussed Benny's dilemma, resulting in her suggestion that we have lunch with George Balanchine. While that lunch did not come to pass, Violette and I did enjoy meeting with Mr. Balanchine in his offices at Lincoln Center. Most graciously, he produced a drawing that depicted the perfect dance floor and gave me a copy. Of course, I dispatched that copy to Benny, who was more than grateful. Later, Violette departed New York for Europe, there to direct the Paris Opera Ballet. Such a talented, lovely, and delightful lady!

Within a short time at Worthington-Skilling, and now barely thirty years old, I became involved almost simultaneously in the design of two innovative buildings for the IBM Corporation: one in Pittsburgh with architects Curtis & Davis and the other in Seattle with Minoru Yamasaki. Both were significant in advancing the concepts of the World Trade Center and the US Steel Tower, in Pittsburgh, leading to an understanding that I could provide significant contributions to a project's overall architectural design.

The Seattle Convention Center interior, before renovation. This space was significantly altered to become a modern facility, the Seattle Concert Hall-Convention Center, for the 1962 World's Fair.

IBM BUILDING, PITTSBURGH, 1960–63

In the late 1950s Curtis and Davis Architects of New Orleans suffered an extensive failure in the patterned hollow tile of their design for an exterior sunscreen. Our John Skilling had met them earlier. Apparently leaving a lasting impression of the quality of our work, they sent to us some photographs asking if we could sort out their problem. After studying the photos plus appropriate drawings, I decided that a relatively simple, quick to construct, and economical solution was available. I was on the next plane to New Orleans, to see firsthand that my proposal was sound, and architecturally acceptable. Indeed, my proposal was heartily embraced by both Buster Curtis and Arthur Davis, though they would rather have seen our John Skilling instead of this less than thirty-year-old engineer.

On one occasion, as Arthur, Buster, and I were enjoying dinner, Arthur ordered a bottle of wine. The waiter asked for proof of my age. The two architects, roaring with laughter, teased me unmercifully for this teenage boy's attempt to consume a glass of wine. Still, the waiter persisted, offering additional avenues for the teasing.

Later, when we received the commission for the 14-story IBM Building in Pittsburgh, I developed the concept, the analyses, and the detailing of the perimeter structure. As I recall, it was Buster Curtis who proposed the idea that the facade contain diamond-shaped windows, an idea that John Skilling modified to incorporate the structure into the mullions. My responsibilities were limited to just the perimeter diagonal structure; others in our firm were responsible for the structural designs of the remainder of the building.

Now known as a *diagrid*, at that time we had no name for the structural system, thinking of it as simply three-dimensional trusswork. Of interest, as late as the preparation of this book, the diagrid structural system of other buildings was thought of by some as new, more than fifty years after we developed it in Pittsburgh. My innovation, to the degree one existed, was the prefabrication of the structural steel forming the perimeter bearing wall and the use of a variety of strengths of the structural steelwork so as to direct the loads in the perimeter wall in an appropriate manner.

How to analyze and detail the structure? I developed a concept that became one of our signature approaches to structural engineering. The traditional approach had been to establish the yield point of the structural steel, then to determine the member loads and, finally, the member properties taken from the given yield point and the determined member loads. Instead, I first determined the required member properties, all based on performance, loading, function, and aesthetic requirements. Only then was the required yield point to be determined. In the traditional approach, the yield point of the steel was initially assumed as a given; in this approach, the yield point was the final answer.

Being in a precomputer era, to accomplish the analyses I made use of an iterative process, retaining a group of ladies, each with a desk-top calculator. With this approach, and limiting the maximum structural member to pairs of 6-inch by 6-inch angles, I found the need for steels ranging in yield points from the low end (36ksi) to very high pressures (100ksi). This use of the highest strength steel available, a quenched and tempered steel was to the best of my knowledge a first in building construction. The computational method for determining the properties of the structural members was one that I had developed, but there being nothing that is

truly new in structural engineering, it is not unlikely that others had preceded me.

The fabricator/erector, the American Bridge Division of the United States Steel Corporation, came up with the idea of painting each grade of steel a different color (neutral for the lowest grade, blue for intermediate grades, and red for the quenched and tempered steels), thus displaying to the world that the stresses were largest in the lower portions of the structure—all while promoting their own products.

It was early in the 1960s when the temporary timber shoring, supporting the exterior frame, was removed. While there was no question in my own mind, there were rumors that the structural system was unsafe and that, with the removal of the shoring, it would collapse. After all, how could such a big building be supported on 6-inch angles? At the appointed hour, the area was filled with press photographers, all eyes and lenses focused on our structure. When there was no sign of movement as the temporary shores were removed, the (perhaps disappointed) crowds dispersed.

Top left The steel frame of the IBM Building was painted to reveal differing stresses and grades of steel. The innovative diagrid, both skin and structure, enabled column-free interiors and a new typology of tall buildings with concrete cores and load-bearing perimeter frames.

Above The building is basically a vertical cantilever that rests on only two points of support on each facade.

In his book *Life and Architecture in Pittsburgh*, architectural historian James Van Trump, in writing about the structure, made us wrinkle our brows before chuckling:

> There is no doubt that it is a really engaging 'stunt' like one of those old vaudeville 'turns' in which a whole pyramid of acrobats is supported on the shoulders of but one man. We might conceive of the building as a great ornamental lantern that might be any size you please, with its present dimension merely fortuitous. Is this the Pharos of the future? . . . Lighten our darkness, we beseech thee, Oh IBM, and by thy great mercy defend us from all perils and dangers of the night!

Being truly appreciative of any lighthouse that would teach us of the future, it seems to me that Van Trump had thrown his own light so as to clarify our structure.

Carl Condit, the architectural historian, spoke of the design more or less as follows: "A curious form of rigid-frame trusses, are to be found in the exterior walls of the IBM Building, Pittsburgh. These steel lattice trusses bear a surprising resemblance to that invented in 1820 by the architect/engineer, Ithiel Town." While Town's trusses were used in timber covered bridges, the foregoing lends credence to my position that there is nothing really new in the worlds of engineering and architecture.

The IBM Building in Pittsburgh was followed, more or less immediately by the IBM Building in Seattle (see page 103).

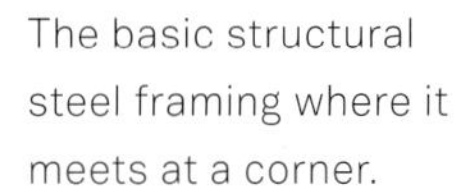
The basic structural steel framing where it meets at a corner.

Above The erection of one of the diagrid panels, representing a major step in the prefabrication of large structural steel panels.

Right One of only eight points on the ground—two on each facade— that support the 14-story building.

APPROACHES TO ARCHITECTURE, ENGINEERING, AND LIFE

ACTIVISM

Beyond the exploring of some of the joys and the sorrows inherent in our professions, in writing this book it is my hope that others, particularly young students of architecture and of engineering, will take from the reading something of the fervor for my underlying beliefs and my need to seek fundamental changes in our world.

As I mentioned, at UC Berkeley I learned the word *pacifist*, though World War II had already taught me that I was one. It seemed that everyone I knew at Berkeley was either a *pacifist*, an antiwar militant, or both. Of interest to me was that this commonality did not seem so prevalent among the students of architecture or of engineering. Perhaps they were more committed to their educational goals, while I was more interested in stopping the ongoing conflicts.

As well, my belief in a benevolent God had been erased by WWII. It was easy at Berkeley to search for something on which to build a philosophic base, this by attending religious services and meetings and by talking with brilliant students of all faiths. I attended services for Buddhists, Christians, Hindus, Jews, Muslims, and others. My appreciation of the beauty of all these great religions came into being; however, despite a concerted effort, none of what I learned took hold of my heart. I remain, today, as one who is not an atheist, one who cannot deny the existence of a God, but one who cannot find a home for his muddled beliefs. Perhaps I just drifted too far from the shore.

Issues of world peace, racial equality, human and women's rights continued to be the guide dog that drove my everyday life. During the 1960s, the Kennedy family mesmerized my political thinking. There is nothing unusual here as all of my friends were like-minded. The assassination of President Kennedy followed by that of Senator Robert Kennedy was devastating. The President was, for the likes of me, too exalted to be approachable. Bobby Kennedy was almost equally distant. Even so, I did stand in that long line at Saint Patrick's Cathedral in New York, but just before reaching Bobby's coffin, I turned away, unwilling to face the reality of his death. The entire affair, St. Patrick's Cathedral and all, now blurs in my mind; I was probably too emotionally involved to see the events clearly.

With an unknown kindred spirit at a peace march in Washington DC.

With the onset of the Korean War in 1950, rising support for racial equality, human rights, and women's rights, as well as the US Senate race in California between Richard Nixon and Helen Gahagan Douglas, I became a minor leader and activist in student uprisings.

I worked in support of Ms. Douglas, not so much because of the positive voice that she represented but because of Nixon's slanderous, sexist attacks. Nixon prevailed.

Similarly, I worked against Senator Joseph McCarthy, who in the early 1950s unleashed unbridled hatred onto the United States. While I had no access to television, his speeches were broadcast over the radio. Being heavily biased against McCarthy's stated beliefs, I delved into his background and was thus able to speak to groups of the untruths that he was heaping upon us. Alas, I was preaching to the choir as my audiences were believers, with only rare skeptics.

It was in this era that I changed from being a pacifist to being an activist-pacifist. Yes, we had our scuffles with the police and, yes, I was arrested many times. While this business of being arrested sounds very serious, it was little more than being driven to the police station, documented, and then released.

One of my Berkeley friends, an ex-Marine officer, elected to volunteer for the Korean War; like me, being a World War II veteran, he was exempt from the draft. To no avail, I sat with him, pleading that he not volunteer. Like so many others, he perished in the Korean conflict. This loss further solidified my anti-war beliefs.

As the years went on, I became increasingly involved in work: seven days a week, many hours each day. Beginning in the mid-1960s, totally engulfed in the design and construction of the World Trade Center in New York and in the United States Steel Building in Pittsburgh, I was not able to join so many of the historic events in the civil rights movement. While I did manage to sneak away for few days to join the huge crowds for the March on Washington for Jobs and Freedom in 1963, I was a bit overwhelmed by the hundreds of thousands of fellow protesters. I'm afraid that, as was the case for so many of my activities as a protester, I had little or no impact on the outcome.

Still, I have tried and continue to try. For example: for various protests in Washington DC, SawTeen and I rented one or more 55-person busses, rounding up friends, students, and compatriots to fill them. Indeed, we encouraged those who eschewed our beliefs to join our bus so as to be able to present their opposing view. For such activities in New York City, I sometimes went alone and sometimes with SawTeen and friends. On one such occasion, following a rally protesting the role of the New York City Police in limiting the activity of protesters, there was a march from Central Park, turning south on Park Avenue and back up north along Fifth Avenue. Bernard Rice (a good friend who played a leading role on I. M. Pei's Bank of China Tower), SawTeen, and I joined the march leaders in the front. Finally, the police brought our progress to a halt, just adjacent to the residence of Joan and Harry Cobb (partner at Pei Cobb Freed & Partners). We waved to them but our faces were lost in the crowd.

The end point of our march was to have been Mayor Michael Bloomberg's apartment, just off Fifth Avenue, but clearly the police would not allow protesters to reach that destination. Pretending that we were somehow permitted to do so, avoiding arrest, we ducked under the barricades and proceeded to the mayor's abode. There, we learned that His Honor was not home, that our message of protest could not be delivered in person. Waving to the assembled press, the three of us departed. In the morning following, since we were the only protesters to reach the mayor's home, we found our pictures in the papers and on the TV as "the protesters."

THE NEED FOR CHANGE

We can change the world. Yes, we can do it! Yes, we must do it. Perhaps, before setting out to make change, we should discuss first the changes that need to be made.

First, a taste of history: It was President Teddy Roosevelt who dispatched the Great White Fleet around the world, setting the stage for the United States as a military and economic power. The carnage of World War I was shortened by the intervention of our military forces, but later Franklin Roosevelt brought the United States into the wars against Germany, Italy, and Japan. Then, there were the conflicts in Korea and Vietnam, followed by our continuing military actions in the Middle East. For the most of my life, our world has been filled with military conflicts.

For some reason, the root causes of war can be explained in so many "nice" ways. Have we killed almost 5,000 young American soldiers (more than two for each and every day) and more than 100,000 Iraqis, all because of our fear of undiscovered weapons of mass destruction? Or is it, as others claim, that the war is a way of a son or a daughter, be they politicians or other, to prove themselves to their fathers? WMD has become an all-too-familiar acronym.

What about the Syrian refugees and the million or so slain in the African civil wars? Being far away, it is so easy to forget them.

But let us not forget the past: the burning of heretics or the *auto-da-fé* under the Spanish Inquisition; the tens of thousands of Protestants in the Spanish Netherlands, hanged under the iron fist of the Duke of Alba; the Holocaust in Germany; the United States bombs dropped on the civilians of Dresden, Tokyo, Hiroshima, and Nagasaki, and so many other nonmilitary targets—all heinous acts of "civilized" people. Were the perpetrators terrorists? Are we, you and I, terrorists?

So many wars! So many persons killed or maimed for reasons not understandable to the most of us because of the shortness of our memories and our desperate need to forget, wars to be found only in the blurred pages of history books.

Is there not a need for change?

Beyond these wars, why is it that the worst storm clouds in history continue while the ravages of poverty go on, seemingly without abatement? Those of us who have found ourselves witnessing the agony of so many of the peoples of the world cannot question the need for change. The squalor and despair in our own inner cities cannot be compared to the hopelessness to be found in Palestinian refugee camps, in the *favelas* of Rio de Janeiro, in the hundreds of thousands of towns and cities of Africa, in AIDS clinics across the world, and on, and on, and on.

Any person in any country of the world can find himself facing abject poverty within just a few miles of his residence. And yet, we—yes, you and I—are a bit overweight from our too-rich diet and our easy access to health care. Indeed, it is only when we move outside of our daily routines that we become aware of others who are less fortunate. Far, far less fortunate. Hey, most of our pets, your cat and my puppy, live a better life than do the peoples of the Third World; the pet food sections of the local super market stretch on and on and on.

With fellow protestors in Washington, who arrived on our chartered buses from New York in junkets of 55 or 110 persons.

Is there not a need for change?

The irresistible force of the population explosion in the developing world contributes to the creation of poverty, hunger, and sufferings that are beyond our imagination. Why did Presidents Reagan, Bush, and Bush refuse to provide funding for education and medical advice related to birth control? Are we to consider these perpetrators of violence to be "terrorists" because they have denied human dignity to millions of Africans?

Over my lifetime (1928 to 2017, so far) world population has increased from about 2 billion to more than 7 billion—more than 350 percent. Should I live to be a hundred, it is anticipated that the world population will increase by another 1 billion, to more than 9 billion. Should we reduce energy consumption per capita in half, does not this heroic effort in reducing greenhouse gasses disappear where the population doubles? Is not the war on climate change intimately tied to the right of women and men to choose or not to have a baby?

What if just a few visionary men and women were in the right place at the right time? Can change be made possible? Had Al Gore, who had the most popular votes, become president, the Kyoto Protocol might have been inked by the United States, an International Legal Court might have been settled, 9/11 might not have taken place, and a second Iraq War would be almost unthinkable.

Instead of those who perished on the 11th of September becoming the guiding star for global responsibility, we are now seen worldwide as arrogant war mongers. How could we have allowed our government to put us into ever-increasing peril from those who see us as money-grubbing, gun-toting ignoramuses? Yes, just a few visionary men and women, at the right place and at the right time, could turn our planet around, set it on a new course of peace, freedom, education, good health, love, and prosperity for all.

Is it not true that 9/11, being a brazen attack on the perceived symbol of capitalism, would not have taken place had the United States turned its wealth and its power toward bettering the conditions in the developing world? Is it possible to understand how a young woman or girl would tie a bomb to her body when she and those around her can join the rest of us on the road to equality? No, it cannot be comprehended, and it would not happen.

But there is another side to the story, one not always for the better. About twenty-five years ago I was asked to make a short speech before *Engineering News-Record's* (*ENR*) black-tie annual dinner; my audience was well-filled with important persons from the construction industry. I proceeded:

> Somehow, we need to recognize that we can make a difference. Where our hearts and our energies are sufficiently stalwart, we will make a difference. Let me cite just a few of the many thoughts worthy of consideration:
>
> - How are we to build the kind of world that our children deserve? What can you and I do to make a difference?
> - There is this draining of our energies from building and from life toward destruction and death. Aren't we still concerned with the "guns versus shelter" argument?
> - The built environment should be redirected along more humanistic lines.
> - An important government official said that you get the best bang for the buck from nuclear weapons. Is that thought still alive?
> - My God! The high school drop-out rate is approaching one-third and many of the graduates can't read and have no skills. We need to better educate kids for the construction industry!
> - And I've saved one of the most provocative thoughts for last: What about a high school for the construction arts?
>
> How about that? A high school devoted to the needs of the construction industry; quoting *ENR*, the construction industry is the largest employer in this nation!

The members of ADPSR (Architects, Designers, Planners, for Social Responsibility) are not flaming liberals nor straight-laced conservatives. Instead, they are people like you and me who believe in the basic values of these United States of America, folks like Bruce Fowle, Sidney Gilbert, Jordan Gruzen, Jim Polshek, Joel Silveman, John Tishman, and many others. Each of us in ADPSR aspires to be one of the president's 'thousand points of light.' Each of us recognizes that the great groundswells of social change are not created by large organizations but by smaller groups of thoughtful, committed, and hard-working citizens.

An *ENR* editorial pointed out that each of the accomplishments of the construction industry is but one chapter of an illustrious history of human achievement and welfare. It stated that this is so because the community of architects, engineers, and contractors has the potential to produce a world of unparalleled beauty, safety, and hope.

Well, I believe in the basic precepts of that editorial, and I know that you believe in them as well. Whether it is a high school for the construction arts, affordable housing, day care, disarmament, drugs or other issues, such ideas do not reach fruition in the absence of creative and dedicated support, support from folks like you and me.

Please, reach out! Get behind the construction industry by bringing your wisdom, your intellect, your energy, and yes, your pocketbook to bear on the needs of our peoples.

In preparing my remarks, I had no intention of soliciting funds nor the efforts of individuals. Instead, my goal was to remind them that in the construction industry, their industry and mine, there were far-seeing persons who had not forgotten their years at the university, years for banner-waving, years for reaching out to those far less fortunate than the least wealthy of those in the audience. Measured one way, I failed, for not one single person as much as sent me a post-card. Perhaps my heartfelt words had fallen on deaf ears.

Yes, there is the need for change! To evoke any such change, all of us need get out from behind our daily duties to raise our voices on behalf of all on this planet. I am proud to report that the inbox of my computer is constantly filled with issues of human and women's rights, freedoms, and the like. There is no end for the need for each of us to become involved in issues of social activism. If you have not done so, please join us!

The Origins of the World as Seen Through the Eyes of the Structural Engineer

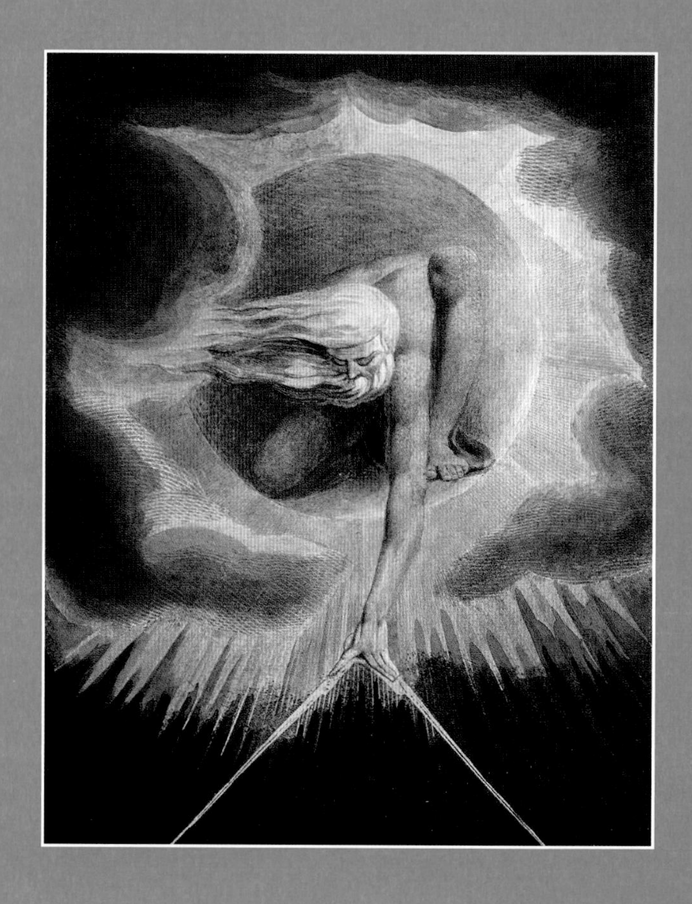

William Blake, *The Ancient of Days setting a Compass to the Earth*, 1794.

IN THE BEGINNING...

On the first day, God turned to the angels, "I wish to create a universe. What should be my first step?" The angels responded more or less in unison: "We need architects."

And so it came to pass that God created architects, dressing them in black jackets and soiled and wrinkled black pants with red sox and sneakers. He blessed them with imagination, creativity, artistry and professionalism, and with some, a complete lack of business acumen.

God was pleased with his new architects. "I've decided to create a universe," He said to them, "and I've made you to assist me."

"Good choice!" said the architects, "now we will need a full team of engineers." "Why?" asked God, puzzled. "All of the boring bits," the architects thought to themselves, but they answered, "All of the specialized structural designs, building services, and the like."

"You mean all of the boring bits," said God, perceptively.

There was much shuffling of feet and down casting of eyes, but in the end the architects received the engineers that they needed, dressing them in tweed jackets and blessing them with ingenuity, creativity, professionalism, bravery, and leadership.

With the project team now ready, the architects decided to call the first meeting to discuss the problems at hand. "How many chairs will you need?" asked God. "Can't anyone here count?" The architects demurred. "Don't ask us," said the engineers, "we don't count either." God was perplexed.

"It appears that we need yet one more consultant, someone to count," reasoned God. The project team nodded in full agreement.

"What shall we call him? The counter of chairs?" asked God?

The consultants deemed that the proposed title was a little demeaning. The angels suggested: "Noter of the Numerological, Manager of the Mathematical and Assessor of the Arithmetical."

God frowned.

"What about the Estimating Surveyor of Quantities?" suggested the structural engineers, shyly, because engineers were not supposed to have verbal or literary skills.

"Perfect," said God. "I shall create a quantity surveyor."

And so God, perhaps not realizing the full significance of his act, created the quantity surveyor who was pragmatic, reliable, suitably conservative, and with a good mind for numbers, balanced by a total lack of aesthetic or philosophic concern. Also, God gave him all of the bits of wisdom and knowledge left over from the architects and the engineers (which wasn't much). The structural engineers were allowed to give the quantity surveyor a nickname: 'The QS'.

Unfortunately, having created so many architects and engineers, God didn't have a suitable persona remaining in his warehouse. All that was left consisted of some pin-stripe suits and bowler hats.

Unperturbed by an image not entirely consistent with his exalted image of self, this new personality stepped out among the assembled consultants. The quantity surveyor's first task was to count the number of chairs required, and he got that right (actually, his count was within ten percent, which he believed to be quite close considering market factors, escalation, unforeseen items, and contingencies).

Since there was not all that much for him to do, the QS sat in a quiet corner and listened. The architects set out to design the Earth and, Oh! It was soooo beautiful! The earth was to have lots of curves, angles, and squiggly bits, and glass mountains; textured renderings, all pastel, added the magical finishing touches. The structural engineers designed the support system: very sturdy, very safe.

The mechanical and electrical engineers devised an ingenious sun in the middle of the sky to make it warm and bright, a smaller yellow sun in the east for the cool morning, a red sun in the west for the evening, and a soft gray-blue one in the night so that folks could sleep. The architects added lots of stars for dramatic effect and God was pleased.

"Who shall build this wonderful universe?" God wondered. And so He created the builders, asking them how long they expected to take in construction. Following forty days and forty nights of study and deliberation, the builders returned with their quote: "Not seven days, not seven weeks, but seven times seven weeks," said the builders, "and we must have 30,000 pieces of silver."

God was shocked. "But I must have the universe in six days and for thirty pieces of silver."

God called the consultants together for yet another meeting (there were almost enough chairs to go around), telling them that He was concerned that the universe was going to take too long to construct and that the cost was substantially over the budget. "Terrible, that the cost of building a universe is out of reach these days," muttered the architects.

"Good design may take a little longer to construct," opined the engineers. There was much rhetoric, shaking of heads, and commiserating. God was about to abandon the project.

"It could be done quicker and cheaper," said the QS in a resonant voice from his dark corner of the room. "A few changes could reduce significantly both the time and the cost of construction."

"How would you know?" the consultants asked, "You know nothing about the construction of universes."

God turned to the quantity surveyor, advising him that he had been created only to count chairs. God added that He was well pleased with that initial effort, while recalling that the quantity survey had got it right within ten percent, anyway. Then, on reflection and in a move that has since plagued the construction industry, God, looking down on the quantity surveyor, asked: "How can I build this universe more quickly and at a lesser cost?"

"The most important thing," said the quantity surveyor, "is to make the earth round. All of those compound curves, angles, and squiggly bits are too expensive."

The structural engineers, warming to the battle, said (but still speaking shyly so as to not offend the architects): "Let's do away with the foundations. If you, God, create gravity and orbit the Earth about the sun, it will stay there without any foundations (structural engineers being very wise in such matters). Also, you don't need the smaller suns for morning, evening and night," the structural engineer proffered with an air of authority.

"Yeah, right" said the services engineers, "It'll be broad daylight all of the time so that no one will be able to sleep."

"Not if we spin the earth on an axis," said the structural engineers. "Then, the sun will not be shining brightly all of the time. Also, a small moon to reflect the light of the sun will be more energy-conserving and will cost less."

The QS, jumping into the battle added: "Why not create more architects, engineers, and builders so as to increase competition and reduce fees?"—wisely omitting quantity surveyors from his list.

God agreed, creating hosts of architects, engineers, and more builders, and being nobody's fool, added one or two quantity surveyors as well.

With much weeping, complaining, and gnashing of teeth, the enlarged cadre of architects changed the design. They presented their revised plans to Him for acceptance and approval (withholding warrantees, expressed or implied, and leaving the most of the detailing to the builders).

God then asked the quantity survey if he could count things other than chairs. "Of course!" said the QS, "and I'm particularly good at beans."

"Good!" said God. "Can you tell me how long it will take to design and to build this new universe and, naturally, how much it will cost?"

The QS responded: "Based on the scanty information made available by the consultants, and taking advantage of the reduced fees provided to the architects and to the engineers, along with the exclusion of contingencies, escalation, market factors and the like, my non-guaranteed budget estimate is seven days and seven nights plus thirty-five pieces of silver."

"Great!" said God, "That's close enough to my budget."

Following another forty days and forty nights, the cadre of builders returned with their lowest and final offer: "Thirty pieces of silver and seven days and six nights, but you must strike the liquidated damages clause and allow for price escalation, all as provided to us by the QS."

God, being reasonably happy with the builders, signed the contract.

The quantity surveyor was delighted. "At last," he said, "I have proven my worth."

But the consultants were not happy.

"You have ruined my design!" shouted the architects. "You should leave well enough alone!" complained the engineers. It should be noted that both were far too professional to mention the reduction in their fees associated with the lowering of the cost of construction and with the increased competition.

The quantity surveyor turned to God for condolences. Surely, God would be happy, what with all of the savings in time and in money. God took the quantity surveyor aside to tell him that his estimate was wrong by fourteen percent, chastising him for not measuring up to his title as a Surveyor of Quantities. Indeed, God was not completely happy on account of the allowed changes to the contract. He banished the QS to a mansion in the highland overlooking the hovels occupied by the engineers and the brick houses with two-car garages inhabited by the architects.

In his mansion, the QS was content to entertain such accountants and lawyers as would condescend to visit with him. The QS continues to count chairs and now, also counts bricks, doors, windows, and other items uncountable by architects and by engineers. He is completely happy, though believing always that God should be ever more bountiful toward him.

Leaping above the clouds on the slopes of Mount Haleakala, Hawaii.

SOME VIEWS OF ARCHITECTURE AND ENGINEERING

DEFINITIONS

The dictum that architecture is accepted as central to the liberal arts (which it is), while engineering has been deemed irrelevant (which it is not), is silly in today's world. Underlying this misconception is the notion that engineering is a purely rational activity; that, for each technical problem, there exists, as Jacques Ellul put it in *The Technological Society*, a single solution that is the "one best way. All that engineers have to do is to find it."

In his book, *A Life in Architecture*, Minoru Yamasaki wrote:

> Yet, in the end, it is architecture, not engineering, that creates the basic concepts for a building. One can take any particular engineering firm and, looking at the various projects on which it has cooperated with different architects, find a wide range in quality in the structural imagination exercised in each separate building. This is because it is the architect who, through their design concepts, set the standards for the building. Thus the engineering firm, no matter how good it may be, can only go so far in producing an appropriate and exciting building, for much of what it accomplishes is predetermined by what the architect has requested.

He then went on to discuss a variety of structural systems and concepts. Of some interest, in his book, perhaps fearful of omitting anyone, Yamasaki does not recognize any of the many talented persons who contributed to his successes. Nor does he recognize the introduction of architectural forms inserted by many engineers into some of the most important buildings of our time.

> Technology seems to dictate that artistic expressions or personal tastes are "frills" that engineers must do without. This is the logical outcome of the "applied science" view of technology. Blind to the engineer's aesthetic and societal achievements, and ability to create both humane and monumental buildings and structures, many writers see only the march of logic and efficiency, trampling art, sensitivity and humanity itself.

Of possible interest, my professional license in Japan is as Architect and Building Engineer.

Some years ago, Aldous Huxley, scion of a family of distinguished scientists and author of *Brave New World* (1932), asserted that modern "technological systems of production and organization are virtually fool-proof." He added: "If anything is fool-proof, it is also spontaneity-proof, and inspiration-proof." Advocates of the "one best way" must assume that an optimum can always be found if we have computers that are powerful enough and analysts who are clever enough. But, to me, that is an illusion. There are altogether too many basic qualitative measures of design to allow the existence of such a "best way."

In 2013, the architectural critic Martin Filler, wrote:

> Despite the present image of the architect as the heroic loner erecting monumental edifices through sheer force of will, the building art has always been a highly cooperative enterprise. Although the *parti* (basic organizing principle) of a design may sometimes be the product of just one intelligence, the realization of a structure of even moderate complexity depends on a broad range of expertise seldom encompassed by any individual, no matter how singularly gifted. As an artistic endeavor, present-day architecture most closely resembles filmmaking, in which the prime creative mover, the director—even the most visionary of auteurs—requires the specialized technical skills of a large cohort of indispensable collaborators.

The reader is referred to another part of this book wherein is described the largest privately owned building in the world, which owes its plan shape and its exterior aesthetic to the structural engineer—in part refuting the "one intelligence" argument. (See US Steel Tower, p.173)

Taking another view, and quoting from the writings of Ada Louise Huxtable in the *New York Times* in 2004:

> The time is past when the architect was the form-giver who handed an idea to the engineer to make it stand up. Today's structural engineer is a coequal designer. But names like Cecil Balmond, Guy Nordenson, and Leslie Robertson are virtually unknown outside of the profession.

Is it possible that Mr. Huxley, Mr. Filler, and Ms. Huxtable are all correct in their views of the engineer? That is, perhaps every architect is part engineer and every engineer is part architect, with some architects being more engineer than architect and some engineers being more architect than engineer?

In seeking definitions of *architect* and of *engineer*, I ran out of patience and settled for less than I'd hoped. As an introduction, it should be said that it's impossible to know where you're going when you don't know where you've been. Accordingly, we need turn to the past.

In the middle of the nineteenth century, the Institution of Civil Engineers of the UK asked the practitioner Thomas Tredgold to prepare a definition of the term *civil engineer*. Recognizing that the term denoted any civil engineer, including those not engaged specifically in the works of the military, Mr. Tredgold penned the following:

> The art of directing the great sources of power in nature for the use and the convenience of Man.

What a fine definition!—one used by the Institution to this very day.

This I'm told, but find it hard to believe, is taken from the Scottish branch of the British Institution of Structural Engineers:

> Structural engineering is the art of modeling materials we do not wholly understand into shapes we cannot precisely analyze so as to withstand forces we cannot properly assess in such a way that the public at large has no reason to suspect the extent of our ignorance.

Seeking further "definitions," with the *Encylopaedia Britannica* being a likely prospect, I thought that I could go back to the first edition; it was printed in 1771. Approaching the librarian in the Rare Books Department of the New York Public Library was a bit like approaching a snowman. She said that I was too clumsy and, besides, the eighteenth-century publication was in a vacuum or something. Still, she did read the definitions to me. The second edition was printed in 1779 and the third in 1797, both containing language identical to that found in the first edition. Presumably, there were no yearbooks or whatever. Life was a bit slower then. In any event, the definitions read as follows:

> ARCHITECT: A person skilled in architecture, or the art of building; who forms plans and designs for edifices, conducts the work, and directs the several artificers employed in it.

The word is derived from "workman," the "principal workman."

> ENGINEER: One in the military art, an able expert man, who by perfect knowledge in mathematics delineates upon paper or makes marks upon the ground describing all sorts of facts and other works proper for offense and defense. Engineers are extremely necessary for this purpose; wherefore it is requisite that, besides being ingenious, they should be brave in proportion.

In truth, those design professionals (be they architects or engineers) who sail in the shark-filled sea of professional indemnity must be "brave in proportion."

While it's important to recall that the roots of civil and of structural engineering are founded firmly in the soil of the military, we need recognize and respect that our future lies equally firmly in the need to guide our country along the paths of freedom, and of dignity, and of peace, and of equality.

Finally, Professor David Billington expressed the thought as well as anyone:

> Science is discovery;
> Engineering is design.
> Scientists study the natural;
> Engineers create the artificial.
> Scientists create general theories out of observed data;
> Engineers make things, using often only approximate theories.

Engineers who see their profession as an applied science fail often to look back. Yes, it is my experience that those architects and engineers who are the creators and the designers of large-scale works need understand that they are not applied scientists and that they must look back, or their creations will be mundane and they will not long survive.

For our concluding definition, quoting Richard N. White, the (now deceased) eminent Professor from Cornell University, we turn to the older Yellow Pages of the London telephone directory. Under the heading of "Boring" we are directed:

"See Civil Engineer."

Archimedes lifting the world with a lever, 1824.

RECOGNITION & RESPECT

Some years ago, taking part in an American Society of Civil Engineers (ASCE) symposium, one of the topics was the recognition, or the lack thereof, of engineers by the public. Stated broadly, my position was that engineers are recognized or not by society because they do or do not overtly participate in the lives of those around them. In short, one cannot expect to be recognized where one cannot be seen.

During the question & answer session following my lecture, one noted professor challenged me along these lines: "Look, I'm a very busy person, without time to have a public presence. All I want to do is my research, and I expect to be respected for the work that I do."

I refrained from pointing out that his role as "professor" carries a modicum of responsibility to transfer his knowledge to others, i.e., to teach students, not just conduct research. Of course, like others, he is respected by an inner circle of colleagues familiar with his work; however, in essence, he wanted to be both respected and recognized for his title—Professor—not for the contributions that he makes to society.

It was Albert Einstein who counseled:

> It is not enough that you should understand about applied science in order that your work may increase men's blessings. Concern for the man himself and his fate must always form the chief interest of all technical endeavors; concern for the great unsolved problems of the organization of labor and the distribution of goods in order that the creations of our mind shall be a blessing and not a curse to mankind. Never forget this in the midst of your diagrams and equations.

That professor in my audience is fated, with others like him, to be disappointed in the level of public recognition he receives, not because his work is not important, which it is, but because he cannot be seen. Respect, yes. Recognition, no.

An architect, for example, tends to grapple more directly with many of society's problems and is perceived by the public as one who does so. The structural engineer, in his or her day-in, day-out work effort, may not be so publicly involved and is not as well recognized for his or her contributions to the built environment and to society. This is not because the work is unimportant, for surely it is, but because the work is not perceived as contributing to the society in which people live and work.

"Sure," many will say, "the structural engineer makes an important contribution to society but so does the engineer who designs the machine that made the door knobs essential to the workings of buildings." But, that machine would not exist without the essential contributions of the machinist, the electrician, and so forth. All parties provide an essential service to those using the building, but it is almost always the architect who is credited or criticized for the end product. It is important to recognize that the entire team, through their talents, hard work, and dedication, should be recognized in these group endeavors.

That said, a recognition I feel most comfortable with is not strictly related to our industry but in my participation in social improvement. Within our community, the recognition and respect shown to SawTeen and to me are due in part to our efforts in low-income housing, in the peace movement, in seeking support for women's and human rights, in pro bono work, in various professional organizations (engineering, architectural, and general). Far afield from structural engineering, we try to make our efforts or at least our presence felt. For those activities that are driven by our desire to participate in societal advancement, I do feel that a level of recognition, be it ever so small, is attached thereto.

But it doesn't stop there: measured another way, we lead by example. In 1995, our firm set out to have constructed—pro-bono and using student labor—a skateboard park in the northwest corner of Manhattan. Firms from the construction indus-

With Katie Hill from our office and some of the high school students who constructed Riverside Skate Park in New York City.

try rose to the challenge. With the park to be designed and constructed by students, we needed only money for a couple of shop instructors to provide guidance to the students (provided by government), materials and tools to be donated by contractors, the site from the Parks Department, and local restaurants to provide lunches. One of our young engineers, Katie Hill, trained the students on CAD in order to create the needed drawings and worked countless after-work hours to pull it all together. My point is that we can make it happen, but that, in order to do so, this effort engages people—individuals who are motivated to come together—not audiences. As an aside, at the opening ceremonies and in the press releases, neither Katie nor others in our company who assisted were recognized.

EDUCATION OF THE STRUCTURAL ENGINEER

The following very brief thoughts are adapted loosely from the words of others, with special thanks to the Institution of Structural Engineers (IStructE), to the Royal Institute of British Architects (RIBA), and to Robert A.M. Stern, whose lecture, upon receiving the Vincent Scully Prize in 2008, was inspiring to those of us who attended. While these original sources were directed to the architect, many of my interpretations of these concepts are applicable as well to the structural engineer.

I have taught and lectured at elementary schools, high schools, and at many universities around the world, yet my teaching, focused on concepts of design, is limited to a bit less than ten years. At Princeton University my students were both

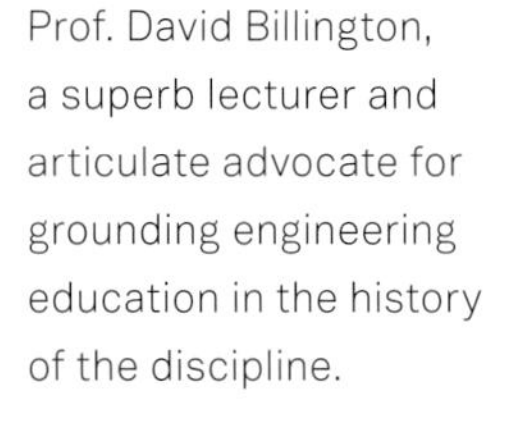

Prof. David Billington, a superb lecturer and articulate advocate for grounding engineering education in the history of the discipline.

undergraduate and postgraduate and from the colleges of both engineering and architecture all mixed together in one classroom. Of some importance, it is essential to understand that I am an educational amateur, however opinionated.

Some argue that engineering education should begin with the history of the discipline. This is fundamental to the wonderful teachings of Professor Billington, but beyond him and some others, history for engineers at the university seems to remain its own pursuit, unrelated, and even disparaged in the morass of technology focused squarely on the future. Stated simplistically, the university should aid both students and professors in becoming as informed as possible in their approach to their profession, and to life, and to inspire an openness of mind and love of learning.

THE COMPUTER

Any discussion of engineering education cannot avoid consideration of certain systemic changes that have come about in the way structural engineers approach their core work. In particular, we should not ignore the fact that computers and their ilk have transformed engineering practice and dramatically ingrained themselves into almost all aspects of engineering education.

Should we not call into question the abandoning of the age-old ways that engineers historically used to develop their insights into the geometries of form and the representation of design ideas? Graphical methods of analyses instill a basic understanding of the structure not provided by computer analyses.

I have no suggestion as to how to entice students and faculty away from their computers. I do know that, while marginally dependent on my own computers, I cannot design without resorting to pencil and paper. I suspect that it is in practice, not at the university, where the young engineer may be able to achieve greater freedom and independence.

Some know me as one of the first to introduce digital techniques into practice. But in welcoming the new digital media, I have not condoned the abandonment of all that went before. The computer seems to foster some structural engineers and architects with little ability to sketch their work, nor do they seem to even understand that they lack this critical skill. While I have no idea of how to do it, professors today should inspire students to use the computer not as a replacement for drawing by hand but as a valuable supplementary tool. Unfortunately, many professors, awash with the glories of the computer, do not have these skills and so cannot teach them. For many architects and engineers, including me, the mind/arm/hand/pencil connection is integral to the creative process. Pencil in hand, design begins on paper, not in the computer, thus capturing and communicating ideas, to yourself as well as to others. To sketch is to analyze. Sketch and estimate before turning to the computer!

Above In 1964, to design the World Trade Center, we established an office in New York equipped with drafting tables, parallel rules, desk calculators, and (not shown) the firm's first computer: the room-size IBM 1620 Data Processing System.

Top right LERA offices in 2015, with smaller, more powerful computers at every workstation.

Today, a young engineer, when faced with even trivial problems, turns immediately to his or her computer, therein finding irrefutable data, whereas my mind embraces a sketch first, analyze later mode. While the computer seems to yield a kind of finality and indisputability, my sketches can be wiped away with the stroke of an eraser, while opening new horizons and new ways to see. Clearly, this is not my field of expertise, yet it seems to me that such successes as I have achieved are founded on the graphics taught to me by Bill Bertwell. My sketch first, analyze later philosophy does not sit well with today's young engineers, who turn immediately to their computers.

A case in point: I set out to give my Princeton students very simple structures, asking them to sketch the deformed shape. While being some of the brightest students in the country, both architectural and engineering students proved to be shockingly inept at doing so. While I had prepared forty odd such structures as homework, after just four, it became appallingly clear to me that the students were simply not able to visualize the deformations, shears, and moments for these very simple structures, but they were fully capable of having the computer determine it for them. Accordingly, disappointed but unable to do more than admonish them, I abandoned this approach.

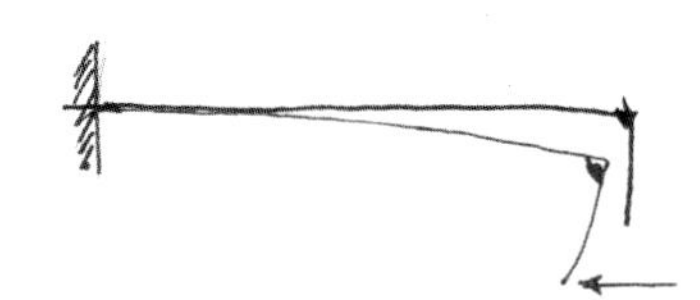

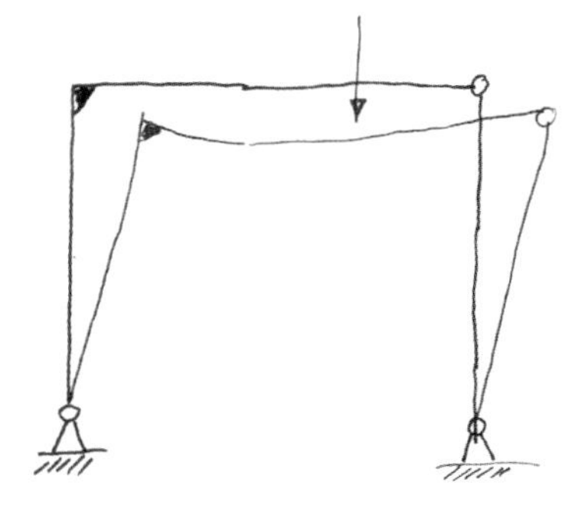

Right The bending shape of these simple structures should be readily understood by students of engineering and architecture.

It seems to me that the heart of this lack of essential understanding lies in the teaching of the analysis of such structures. Shouldn't the teaching of analysis start with the intuitive sketching of the deformed shape, along with the moment and shear diagrams, to be later followed by the analysis? In other words, intuition and sketching first, analysis second. In this way, both the student and the professor may come to a better understanding of something of the structural behavior rather than relying solely on the computer to provide a superficial understanding. Yet, all too many students and professors see the computer output solely as moments, shears, stresses, and the like. They somehow miss the conceptual skill that should be, in my view, an essential ingredient in all analysis courses.

While stresses and strains are important, it is the deformations and the motions of structures that establish those structural systems that are at the core of my profession, from the simplest steel beam to the wonderful and complex structural systems that we are able to create.

TRADITIONAL EDUCATION

I've given each of my university students a copy of Strunk and White's lovely little book *The Elements of Style*. The students seem to turn it over in amusement; they are disinclined to raise their voices in admiration for the clarity of the thoughts found therein. Yet it seems to me that in their ability to communicate, undergraduate students of both architecture and engineering who are skilled in writing and in the arts rise above other, less well-rounded classmates.

I continue to believe that both students and professors should get out from behind their computer screens so as to participate more directly and more fully in the world of social change, peace, civil and women's rights, the environment, and other important causes.

In conclusion, it is important to iterate that I am not qualified to evaluate the effectiveness of the educational systems at any university, and that some of the ideas to be found herein are blatantly taken from others—wiser than myself—but they are ideas in which I firmly believe.

The History of Construction As Seen Through the Eyes of the Structural Engineer

2,500,000 B.C. (08:32 AM) Dud, the world's first structural engineer, piled two stones upon a third, making it possible to reach a nest so as to steal the eggs of an unsuspecting bird.

2,500,000 B.C. (09:32 AM) Exactly one hour later, Wok, representing Cave Man's Local 742, appeared to claim that the structure was not accomplished by union labor.

2,500,000 B.C. (10:11 AM) Wink, who lived just down the street from Dud, while watching the bathing beauties next door, tripped 'n fell over the rocks.

2,500,000 B.C. (10:12 AM) One minute later, Cho (the hunchback from the law firm of Chase & Cheat Ltd.) arrived with the appropriate forms for the contingency lawsuit and for the workers compensation claims.

2,500,000 B.C. (10:58 AM) Unk, wandering into the job site with a hyper sense of authority, hands out astronomical fines for safety violations. Unk is greeted with (and subsequently known as) Oh! Sh!!. Over the centuries, this label has evolved to become OSHA.

2,700 B.C. The great pyramids of Giza are constructed. Designed originally as large cubic structures, Egyptian contractors, faced with a stone-cutters strike, ran out of product shortly after commencing construction. A decision is made to taper the profile as the structures grow, leading eventually to a point at the top. Engineers assert that no one will notice.

2,550 B.C. Along came the first recorded example of "value engineering." With the goal of saving money, Egyptian architects leave off the nose of the Great Sphinx of Giza.

150 B.C. Being quick learners, Greek builders turn to the lessons learned earlier by the Egyptians. In a new attempt at value engineering, the arms are left off of Venus de Milo.

738 A.D. to 1211 A.D. There is no recordable construction in this period. The AIA cites as the underlying cause, the high level of engineering fees.

1370 A.D. The great tower is erected in Pisa. Geotechnical engineers blame the structural engineer for the slight slope that refuses to go away. Architects, noting the influx of tourists, claim that the slope was a fundamental part of the original design.

1666 A.D. While budgeted for 1.4 million francs, the French architectural firm of Henny & Penny screw up the plans 'n specs, with German contractors coming in low at 3.2 million over budget. H & P, blaming the structural engineer for the overage, charges its client on an hourly basis for making the changes to correct its own mistakes. Architects from around the world, noting this daring and profitable move, award H & P the Gold Medal.

1803 A.D. The United States, bilking France, purchases all of the land west of the Mississippi. The

US later asserts that it was screwed because of the lack of properly maintained sewer and water laterals. Natch, a law suit is filed and the firm of Chase & Cheat Ltd. is retained to represent the realtor who consummated the sale. It is anticipated that the case will never be settled.

1810 A.D. Schubert fails in his attempt to finish his symphony, blaming the structural engineer for the delay; accordingly, the architect refuses to sign off on the punch list. The work remains unfinished.

1869 A.D. The transcontinental railroad is completed by the driving of the Golden Spike at Promontory Summit, Utah. The spike is later removed (stolen) by the landscape architect while mowing. The entire railroad unzips, requiring that the whole track bed be again laid. The good news is that the architects and contractors are paid a full fee for the construction of the new track.

1906 A.D. A massive earthquake topples four square miles of San Francisco. Structural engineers blame the software supplier. Insurance companies default, claiming that structural engineers should have warned them of the possible loss.

1912 A.D. With a loss of 1,513 lives, an 'unsinkable' British luxury liner goes down in the northeast Atlantic. Marine engineers blame the refrigeration industry for failing to control the icebergs.

1959 A.D. In the historic Kitchen Debate, Richard Nixon confronts USSR leader Khrushchev at a US trade show in Moscow. Pat Nixon ignored the two blockheads, going with hickory raised-panel doors and gold-fleck countertops.

1989 A.D. Following an exhaustive third-party review of AIA contracts and documents, it's officially determined that, by including an indemnity clause in each and every article, the architectural community has achieved its long-sought goal of not being responsible for anything.

2001 A.D. Airplanes are used to attack the World Trade Center and the Pentagon. Lawsuits against the structural engineers are filed by Chase & Cheat Ltd., claiming that all buildings should provide full protection against acts of war. Even the architects for the project find the time to claim against Les.

2005 A.D. Having spent his life savings renting buses to carry like-minded protesters to Washington, Les Robertson elects to seek a grant from the Department of Transportation. The grant is denied on the grounds that protests are un-American.

2016 A.D. It is determined that there is no hope of attaining peace with the Muslim world, regaining our personal freedoms, or of saving the planet from global warming, much less balancing the budget or providing a sensible level of health care. Chase & Cheat Ltd. claims that a minimum bonus (to the Plan Desk Clerk) is $1,000,000.

FINIS

That First Step

Some years ago the aerialist, artist, writer, and Oscar winner (for the 2008 documentary film, *Man on Wire*), Philippe Petit and his friend Kathy O'Donnell came to our offices, later to join SawTeen and me for dinner. Many of you will remember Philippe, a man of many accomplishments, for his walk in 1974 between the two towers of the World Trade Center. In truth, he didn't just walk. For 45 minutes, treading a wire of steel stretched a quarter mile above the streets, he passed over the abyss eight times, dancing, kneeling, laying down, and performing along the way.

Philippe asked me of the most exciting moments in my own career. I sensed instantly that it was the first step into a new world that occupied his mind, and I so responded to him: "The adrenalin rush, the euphoria that comes as the mind first breaks from the past to embark onto a new road into the future is, at least for me, that which makes all of the hard work more than worthwhile."

Philippe seemed to be delighted with my answer, likening it to his own first steps onto the steel filament that is to be his journey. With one foot firmly on the solidity of a roof top and the second on the wire stretching before him, he mounts the wire for the first time, finding therein a commonality between his world as an artist on the high wire, and mine as a designer of structures.

Each of Philippe's new wires differs from those of the past. The visual environment in which it spans is of utmost importance. Additionally, the span, the sag, the mass of the wire, the balancing pole—the dynamics of his world are never the same. Each time, as he strings his wire and plans his performance, he must reach into his past. He must build on his prior experience. Each time there must be that exciting mystery of the new wire, the new performance.

And so it is in our profession of structural engineering. Each first step must begin with one foot founded solidly on the past, while the second stretches out boldly, reaching for new ground. As well, in both Philippe's and my profession, we must rely heavily on our communication skills, and on our appreciation of art and music, in order to move ahead in our own world.

Yes, over the years our teams of engineers have taken many first steps in the design of projects, most bringing to each a personal joy and satisfaction. Somehow, we are able to personalize our own contributions—be they ever so small or ever so large—drawing therefrom immense personal satisfaction. As well, it's wonderful to physically touch, and enter, that which you have designed.

Young engineers will have advantages over those of us who are old or even older. While we are no longer young of age, make no mistake, we are young at heart and of mind. Indeed, young engineers will need be very fast on their feet to stay ahead of us, to beat us in taking those all-important steps into the world of tomorrow. Of course, young architects and engineers, rising to face their own challenges, will surely overtake us oldsters. But we must remember that successes are to be earned; they are not given.

I am happy to share some of my own first steps in the pages of this book, hoping to encourage young design professionals, architects and engineers alike, to search eagerly for a new and better way. Should you not find on the other side something that is truly wonderful, stretch your wire in another direction until, ultimately, you reach that wonderful world of your own designs.

III

Collaborations with Architects

Minoru Yamasaki 1912—1986

Minoru ("Yama") Yamasaki was born in 1912 to emigrant Japanese parents. I was born in 1928 so Yama was sixteen years my senior, and in many ways, my mentor.

When he was a boy, the Yamasaki family was far from rich, living initially in a hillside tenement in Seattle. With racial prejudice common, this was not an easy life for a *nisei*. Still, the resolute Yama pushed through these difficulties with determination, ending up in the College of Architecture at the University of Washington in Seattle.

It was there that Yama first turned to painting, displaying a singular talent in watercolors. Indeed, I have a reproduction of one of his paintings and have always wanted but could never afford an original.

Much of the cost of Yama's university education was earned by working in the canneries in Alaska. The work week varied from 66 to 126 hours, with the pay at about 17 cents per hour, more for work in excess of 66 hours. The stories of this period in his life, the race to be early in the community bath over fears of gonorrhea, the long hours and the brutal work, attested to Yama's dedication to become an architect.

Before abandoning this very sketchy history, I should say that Yama's first wife, who was also his last wife, was a wonderfully charming and delightful person, Teruko Hiroshik Yamasaki by name, but we called her Teri, a talented pianist from the Julliard School of Music. Teri and Yama were married on December 5, 1941, within two days of the Japanese attack on Pearl Harbor. They divorced twenty years later, only to remarry in 1969.

Between his marriages to Teri, Yama married twice again; once to a geisha, and the other bride was one of his nurses. Their names escape my memory and did not find their way into Yama's book *A Life in Architecture* (1979). In my notes, I found these lines from Yama, which so reflect the inner philosophy of this architect. "Architecture is a fascinating and unique profession that, in its ideal, combines function and beauty to create atmospheres in which man can live, work and enjoy."

He went on: "Each building should enhance the lives of the people who enter or see it." And later, "for me, architecture is not a way of life, it is my life."

In some ways, Yama shared a large part of that life with me. Losing confidence in his structural engineers, true to his beliefs in strong loyalty, in the early 1960s, releasing his structural engineers, he asked our John Skilling to take over the structural engineering for all of his existing projects, whether in planning, design, tendering, or construction. For us, this meant long hours, seven days a week.

It is my belief that when Yama did not see you as with him, he saw you as against him. For example, during the course of the design of the World Trade Center, Yama and our John Skilling had a falling out. Yama immediately terminated all contracts with our Seattle offices, to be finished by other engineers. With no understanding of the problem that separated

Yama and John, after a few months, I wrote to Yama, noting that the problem existed with John, not with me, and that it would be an honor to work with him. Almost immediately he wrote, agreeing with my evaluation and asking that I visit with him. Of course, I did so, re-establishing our relations with Yama, but from New York, not Seattle. Over the course of 25 years, I worked with Yama on dozens of projects around the world, right up until his death in 1986. We remained good friends for decades. During his year-long stay in Memorial Sloan Kettering Cancer Center, I visited constantly with him; because of his health, sometimes for only a minute or two.

I first met Minoru Yamasaki when, several years after joining Worthington & Skilling in Seattle, I was asked to escort him around the construction site of the Federal Science Pavilion, which he had designed for the 1962 Seattle World's Fair. For reasons not understood by me, John (Jack) V. Christiansen of our offices, structural designer of the pavilion, was not chosen to do the shepherding. The honor having fallen on me—which, of course, was wonderful—Yama and I went to the contractor's trailer where the General Superintendent, with pride showing clearly on his face, brought out a model of one of the towering arches that had become a symbol of the facility. Yama, at first startled, bowed his head for perhaps a full minute before looking up to say, "I'm sorry, I made a mistake. I must re-design." Having an extremely tight schedule to complete the construction in time for the opening of the fair, the contractor and I reacted in disbelief. It was then that I realized there was a significant misunderstanding. I said: "But Yama, this is not architecture, this is a model depicting the erection sequence to be used for the construction of the arches." Yama's face changed from a sort of gray, lighting up with his wonderful smile; we finished our discussions. Life was good!

Little did I know at the time that this personal connection with Yama, Jack's wonderful designs for the lacey Gothic arches of the pavilion, and the golden tongue of John Skilling would ultimately lead me to the most important project of my career: the World Trade Center in New York.

Right The wonderful pre-cast/prestressed Gothic arches of the Federal Science Pavilion for the 1962 Seattle World's Fair caught the attention of Port Authority, leading ultimately to our commission for the World Trade Center.

Below Temple Beth El, Bloomfield Township, Michigan, 1973, was a forerunner of Meishusama Hall, Japan (page 165).

UNIVERSITY PROPERTIES

IBM Building

SEATTLE, WASHINGTON. 1961

The 20-story IBM Building in Seattle presented the first opportunity for me to work closely with Minoru Yamasaki and with one of his lead designers, the talented Aaron Schreier. With their offices north of Detroit and ours in Seattle, we scheduled a daily teleconference; fortunately, the calls were usually limited to the weekdays.

While basically a reinforced concrete building, the structure for the entire exterior perimeter wall, from the foundations to the roof, is of structural steel. An unusual design parameter was Yama's requirement that the perimeter mullions be closely spaced, all with the goal that occupants would be comfortable standing or sitting right up against the glass exterior. My idea was to bury the perimeter columns within these vertical mullions of the outside wall. While it would be logical to hide rectangular columns within the mullions, with the narrow dimension in the plane of the wall, such steel sections were available in only limited sizes and grades of steel. I opted to go with round pipe columns, making use of the entire "family" of steels: common steel at the top of the building, quenched-and-tempered steels at the bottom, with intermediate grades of steel between. While conventional designs increased the diameter of pipe in the lower floors, my proposal allowed the use of 4-inch pipe columns, with a uniform outside dimension but with increasing wall thickness, for the full height of the building. When Yama elected to go for precast concrete mullions, I suggested that the 4-inch pipe columns be tucked inside of the mullions, working up a system to bond the precast to the steel pipes, all to add robustness to the column system.

The American Concrete Institute (ACI) came to Seattle for its annual convention; beyond presenting a lecture or two, I provided a tour of the construction site of our IBM Building. Interestingly enough, a prominent engineer, noticing those 4-inch pipe columns, asked me: "Why did you omit expansion joints in the mullions?" As it turned out, thinking of them as mullions, the use of the 4-inch pipes as columns was inconceivable to that engineer.; it was inconceivable to me that the concrete floors cantilevered from the services core. From this I learned something of the frailties of our language and to listen with care to the implausible words of others, no matter how learned the speaker.

Another hard lesson arose as, post-construction, I considered the possibility of water trapped within the hollow pipe columns. My logical mind opined that the contractor would never allow the entrance of water into the columns. My suspicious mind, realizing the corrosive potential of entrapped water, insisted that we do a statistical sampling by drilling small holes at the bottom of some of the columns. As water was found, my fears were realized. The contractor drilled all of the columns, draining the lot. Then, each column was injected with a modicum of Portland cement to erase such entrapped water as might exist. Now, at least for my projects, hollow columns are specified to have seal plates at the top end of each and every column.

Later, after I had moved to New York, the IBM project was recognized by the steel industry as the "Best Engineering High-rise, Commercial, Industrial, or Institutional Construction of the Year." The black-tie awards ceremony, held in New York City, cited many products made from steel, from kitchen utensils to earth-moving equipment. Prior to dinner and the ceremony, Yama and I toured the exhibit; as we passed from steel product to steel product he became more and more agitated, his consumption of alcohol increasing with every step. He was upset because he saw his beautiful building being intermixed and compared with items that he perceived as ugly. While he was able to walk to his chair at the podium, following wine with dinner, Yama was in no condition to accept the award. When his turn came, I rushed to the podium to accept on his behalf. Later, I took him to his hotel room.

Unknown, both to me and to those representing the steel industry, the quenched and tempered steel pipes were imported from Japan into Canada before being shipped to Seattle; had this been known, I suspect that another building would have received the steel award.

These views show the only exceptions to the structural materials of this otherwise all-concrete building: precast concrete architectural cladding (but used structurally) covering 6-inch pipe perimeter columns and the stone-clad steel arches supporting them.

World Trade Center

NEW YORK, NEW YORK. 1963

In October of 1962, Minoru Yamasaki was chosen as the architect for the proposed World Trade Center in New York. He came to our John Skilling, suggesting that we attempt to obtain the commission. I can only suspect that he recommended us to The Port of New York Authority (PONYA, renamed Port Authority of New York and New Jersey, PANYNJ, in 1972), the owner/builder of the project. Although Yama's recommendation may have been influential, we were in competition with many fine New York firms, each having significantly more experience in high-rise design than did we. To prepare a proposal, I opened a one-room office a block or so from our regular Seattle offices.

It was an exhilarating time! I would prepare ideas, from time-to-time John Skilling would stop by to critique some of them, and I would move on to other ideas. The World Trade Center project was to be 100 stories (later to become 110 stories), rather higher than my experience in having designed a 20-story building. Latching on to the closely-spaced columns of IBM Seattle, which we were then designing with Yama, and the prefabrication of IBM Pittsburgh, a project we'd begun in 1961, I molded those thoughts for a five-fold increase in height of the building. Yama was interested in our approaches, but, worried of possible adverse feedback from Port Authority, declined to travel to work with us in Seattle (nor allow me to travel to him in Michigan).

Construction of the North Tower began in August 1968; the South Tower commenced in January 1969. For resistance to wind loads, the towers were designed to be dependent on each other, not as stand-alone buildings.

The day arrived for the presentation to the Port Authority Board. While I sat anxiously in the back row, the golden-tongued John Skilling made the presentation. Much to my surprise, he did not focus on the wonderful ideas we had conceived. Instead, realizing our complete lack of high-rise and New York experience (limited to 20 stories), he emphasized our close relationship with Minoru Yamasaki and how we could assist him in making the project more buildable and more economical. Somehow John knew that Port Authority looked with some skepticism on the technical capabilities of Minoru Yamasaki who had no real high-rise experience, and was located in Detroit, a long way from the culture of New York.

In front of a portion of the proximity model used for the wind engineering of the World Trade Center.

For me it was astonishing that John was able to skillfully communicate the essence of our input in assisting the direction of the architectural concept, even while presenting very little of our proposed ideas. He hit the perfect chords. Although we had worked hard in preparing for our interview, toiling countless hours in concept development, drawings, sketches, and presentation boards, we would not have obtained the commission without the presentation skills of our John Skilling.

Hey! We won the commission and John was able to negotiate a contract. A move to New York was demanded by Port Authority. Having the technical skills, particularly the design and mathematical skills, I was the logical, perhaps the only, person in the firm qualified to lead the team. John Skilling, however, wanted a duo, proposing to join an expert engineer, the late Anton Tedesko, an older and more experienced man and the engineer for the Vertical Assembly Building at Cape Canaveral. I refused, stating that I would bow out in favor of Anton. This brinksmanship on my part had nothing to do with the desire to make money, any lack of admiration for the talents of Anton, or any of the usual reasons. Instead, it was just that I had worked hard and had tabled a lot of good ideas for the project and didn't want to have to turn to anyone for their filtering or further development. With some reluctance, John agreed while I, knowing little or nothing about the running of an office, hiring engineers and drafters and others, paying the bills, having no experience in high-rise buildings, and all else that would be required of me, moved ahead as though I knew where I was going.

That settled, I moved to New York and opened an office at 230 Park Avenue, next to Grand Central Terminal. With me from Seattle, the terrific team additionally included Wayne A. Brewer, Paul S.A. Foster, Ernest T. Liu, Jostein Nes, Richard E. Taylor, and, a bit later, E. James (Jim) White. Professor Alan G. Davenport, on sabbatical from the University of Western Ontario, later joined us to head up the wind engineering research group. In New York, I hired the wonderful Elaine A. Castellano, who started as our telephone receptionist and typist and grew to heading our accounting department, and the anxious-to-learn Angel Franqui, who started as our plan desk clerk and retired as a senior drafter. Hiring others as required, we were able to meet our obligations to the project. Although I was the titular leader, the energies and talents of the entire team led to our successes.

Yama, with more leverage and wiser than this engineer, continued to work from his Michigan offices, leaving the day-to-day coordination to Emery Roth & Sons, who provided a steadying hand on the Yamasaki tiller while producing the construction documents. With broad experience in high-rises and in New York City, they were perfect for the project. Emery Roth founded the firm, and made his sons, Richard and Julian, partners prior to WWII. While Julian was not a licensed architect, he was intimately familiar with the New York market and skillfully guided the architectural work of the World Trade Center.

Still, despite my admiration of and good communications with Julian, there was a constant need to coordinate with Yama and with Aaron Schreier, Yama's Project Director, who was my age and shared my goals and project-related ambitions. Despite the constant use of the telephone and of the facsimile machine, there was the need for face-to-face communications, requiring countless one-day trips to Detroit.

With this constant contact, I became skilled at anticipating Yama's thoughts, and at leading him to generate those ideas that I wanted to incorporate into our designs. I learned the importance of being involved at the very earliest of the blossoming of an idea and, where possible, to be the almost invisible generator of many of the ideas of importance to us or to the overall design.

Yamasaki was a fascinating and complex architect, initially designing almost exclusively with little more than descriptive words and a crude sketch before going on to his model shop; then he would critique them saying, "Let's change this," and "Let's make that more slender," and so forth. For me, much more of a sketcher, it took some getting used to.

Back when he was still seeking the commission, Yama had penned a letter to the Port Authority explaining, in part, his use of physical models in design. "If we should be so fortunate as to receive the commission, we would build a large-scale model of the entire neighborhood surrounding your complex in sufficient detail to enable you and us to understand exactly how the various schemes would relate to the surrounding area. We do this for every job that we undertake. Then, using this model, we would try various solutions in block form, working out rough plans, considering aesthetics, utility, and overall economy."

It is interesting that Yamasaki formulated this letter as bringing senior Port Authority personnel into the process. It is my experience, however, that Yama could not be bowed, that he held firmly to his designs, unwilling to concede on any point thought by him as being of consequence. For example, on two occasions known to me, he resigned from the World Trade Center project over a matter of design; the resignations were not accepted.

During the project's development, it was not unusual to receive a box of wooden models depicting a detail, asking for my opinion. Of course, Yama, who was a skilled painter but seemed to have little interest in drawing, had his own model shop, which he kept constantly busy. Both large and tiny models were construced to deal with areas of design (for his own and our use) and for Port Authority presentations. I've been led to believe that all of Port Authority's models were destroyed in the 9/11 disaster. I have donated models to The Museum of the City of New York, to the Bancroft Library of the University of California at Berkeley, and to the Skyscraper Museum.

Totally demanding of one's energies and talents, this was a young person's project; meaning that if you were of an age to have high levels of "experience" (I was 34), you were perhaps too old to keep up with us. Yama, the late Richard T. Baum (Jaros, Baum & Bolles) and Joseph R. Loring (Joseph R. Loring & Associates), being older, were exceptions, but not their lead designers. This was not a place for people who knew where they were going. Instead it was for those of us intent on discovering where to go and who had a profound desire to discover a way that threaded between practical and theoretical goals. We recognized that there was a desire to create a universally recognized symbol, one that was unique to New York. We set out to design a set of modern buildings, standing on the past, but not based solely on it—rather a new step forward in both design and construction, buildings that would make full use of the latest in technologies, even where we had to devise them. Each floor of the two buildings was almost one acre in size, with the North Tower being 6 feet taller than the South Tower.

Working with the Emery Roth firm was always a bit of a mystery, but there was no mystery about their competence or about their diligence in seeking a wonderful World Trade Center! Just a few examples:

Before departing for lunch, the Roths would enjoy a bit of an alcoholic beverage. While I always declined, staff members were never permitted to imbibe nor, in my experience, join us for lunch; Following an argument with their landlord, employees were obliged to bring their own towels; Julian was never averse to demanding a high level of competence from all consultants; Julian was on one of Richard Nixon's List of Enemies, and proudly kept the list posted in his offices. Alas! I couldn't find my name there!

Enumerating all of the innovations incorporated by us into the World Trade Center would be overly long. The following provides a brief description of just a few of the new ideas and practices conceived and developed by our team. Most, if not all, of this technology is now a part of the standard vocabulary of structural engineers.

THE TUBULAR FRAMING SYSTEM WITH OUTRIGGERS

The structural concept for the World Trade Center was that of a "tube," where the perimeter walls resisted all of the lateral forces of wind and earthquake while carrying all of the vertical loads tributary to the perimeter wall. Although we had used closely spaced columns in an earlier building, it was Yamasaki who proposed that we use narrow windows for the two towers. Our contribution was to turn the closely spaced mullions into columns (3-feet-4-inches on center) making them the fundamental lateral-force-resisting system for the buildings. This tubular framing system also precluded the need for the customary 30-foot column spacing in interior

The staggered field splices of the perimeter structural steel columns: 3-story-high, 3-column wall panels connected by 52-inch deep spandrel plates.

areas, resulting in column-free rentable space, a circumstance that pleased both the architect and rental agents.

Much has been written as to who first created and used the tubular framing system. Almost surely, it was either the incredibly creative Fazlur Rahman Khan (of Skidmore, Owings & Merrill) or our team (Skilling, Helle, Christiansen, Robertson)—one or the other, or both at about the same time. It seems to me that it was an idea in need of realization, making it likely that several engineers, more or less simultaneously, came up with the same concept. In any event, I neither lay claim to nor disavow authorship, believing instead that ideas are creatures of their time, not of an individual or a team.

In support of Yama's design I noticed that, during construction, before the windows were installed, some people felt comfortable walking up to the outside wall, placing their hands on the columns to either side, and enjoying the wonderful view. Particularly, where the wind was in their face, they would walk right up to the outside wall. However, should they feel even a trace of pressure from a breeze from behind, they would at least hesitate before walking to within five feet of the wall, and many would not approach the perimeter at all. It has been said that the closely spaced windows were associated with Yamasaki's fear of heights, but in my experience Yama, being born with a titanium spine, was not afraid of anything; however, to confess, I never walked a beam with him.

Recognizing the lack of historical precedent for a building that grew in the course of design to 110 stories, another structural innovation was the outrigger space frame, which we developed in the eventuality of unanticipated problems. These outrigger trusses, installed immediately below the roof, provided additional stiffening and strength, structurally linking the outside walls to the services core. When the need arose to provide for a rooftop TV antenna (360 feet tall), we were able to unveil our design for the outrigger trusses, which safely supported the antenna, distributing the weight and wind-induced forces to all columns in the building, an essential characteristic since much of the structural steel had been fabricated/erected without knowledge or consideration of the antenna. In this way, the combined system provided additional lateral stiffness, redundancy, and toughness. It should be noted that, for the World Trade Center, the outrigger system was carried in my mind, not in the drawings, as a sort of back door to solve the case should some unsuspected issue arise during the construction. The need for the rooftop antenna proved just such an issue. With the possibility of an antenna atop the South Tower, Port Authority elected to construct the outrigger trusses atop both buildings. Today, with so many new buildings reaching to 500 meters (1,640 feet) and higher, outrigger trusses are commonly employed.

Views of the 360-foot transmission tower, under construction and completed, installed atop the North Tower in 1978.

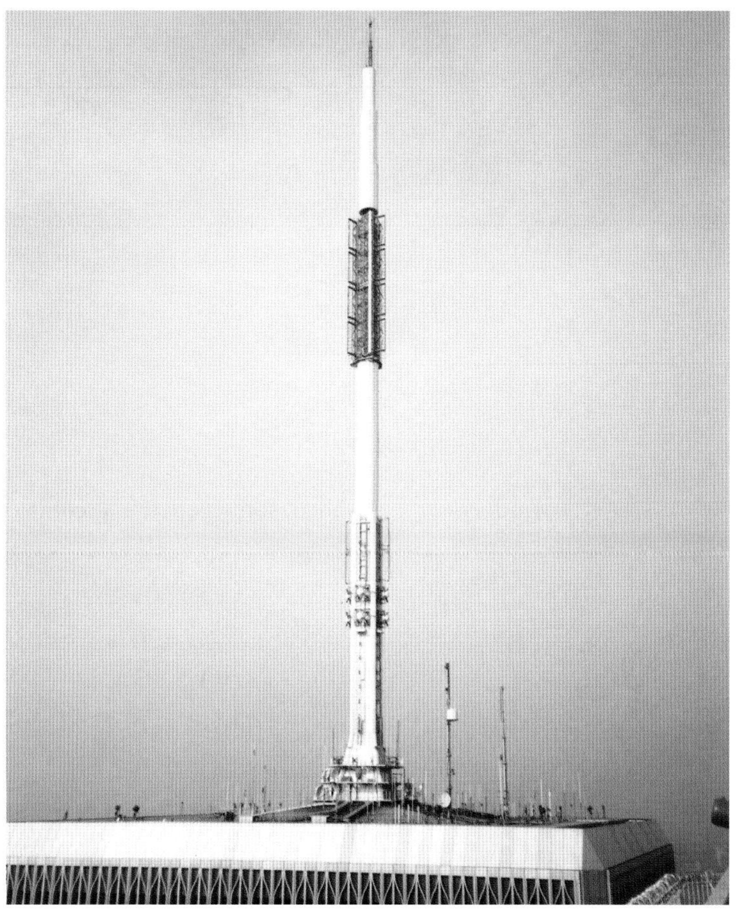

Right and far right Elevation of a typical wall panel accompanied by an array of the data used for fabrication. All of that data was provided to contractors in the form of IBM punch-cards.

Bottom row A host of one-to-three column transfer panels carried on freight cars from Pittsburgh; a "tree column" being erected (note workers for scale); the completed panels installed around the base of the tower.

PANEL	NO.		COLUMN 1 TYPE	COLUMN 1 FY1	COLUMN 1 FY2	COLUMN 2 TYPE	COLUMN 2 FY1	COLUMN 2 FY2	COLUMN 3 TYPE	COLUMN 3 FY1	COLUMN 3 FY2
103	62	59	129	80	65	131	80	60	132	55	55
109	62	59	136	50	50	132	50	50	138	50	50
115	62	59	138	50	50	138	50	50	138	50	50
121	62	59	138	50	50	138	50	50	138	50	50
127	62	59	138	55	55	138	55	55	138	55	55
133	62	59	139	55	55	139	55	55	139	55	55
139	62	59	139	55	55	138	55	55	138	55	55
145	62	59	138	55	55	138	55	55	138	55	55
151	62	59	138	50	50	137	50	50	136	55	55
157	62	59	132	55	55	131	80	60	129	80	80

PANEL	NO.		SPANDREL FLR 62 T4	SPANDREL FLR 62 FY	SPANDREL FLR 62 CONN	REINF. WELD 8 IN LONG COL 1	REINF. WELD 8 IN LONG COL 2	REINF. WELD 8 IN LONG COL 3
103	62	59	11/16	50	343	1·3/16	7/8	7/8
109	62	59	11/16	36	351	1·1/8	1·1/8	1·1/8
115	62	59	11/16	36	349	1·1/8	1·1/8	1·1/8
121	62	59	11/16	36	337	1·1/16	1·1/16	1·1/16
127	62	59	11/16	36	315	1	1	15/16
133	62	59	11/16	42	344	7/8	7/8	7/8
139	62	59	11/16	42	363	15/16	15/16	15/16
145	62	59	11/16	36	368	1·3/16	1·3/16	1·3/16
151	62	59	11/16	36	362	1·3/16	1·3/16	1·1/8
157	62	59	11/16	55	347	15/16	15/16	1·3/16

PANEL	NO.		SPANDREL FLR 61 T4	SPANDREL FLR 61 FY	SPANDREL FLR 61 CONN	REINF. WELD 8 IN LONG COL 1	REINF. WELD 8 IN LONG COL 2	REINF. WELD 8 IN LONG COL 3
103	62	59	11/16	50	348	1·3/16	7/8	7/8
109	62	59	11/16	36	357	1·1/8	1·1/8	1·1/8
115	62	59	11/16	36	355	1·1/8	1·1/8	1·1/8
121	62	59	11/16	36	347	1·1/16	1·1/16	1·1/16
127	62	59	11/16	36	319	1	1	1
133	62	59	11/16	42	348	7/8	7/8	7/8
139	62	59	11/16	42	366	15/16	15/16	15/16
145	62	59	11/16	36	372	1·3/16	1·3/16	1·3/16
151	62	59	11/16	36	365	1·3/16	1·3/16	1·1/8
157	62	59	11/16	55	349	15/16	15/16	1·3/16

PANEL	NO.		SPANDREL FLR 60 T4	SPANDREL FLR 60 FY	SPANDREL FLR 60 CONN	REINF. WELD 8 IN LONG COL 1	REINF. WELD 8 IN LONG COL 2	REINF. WELD 8 IN LONG COL 3
103	62	59	11/16	50	352	1·3/16	15/16	15/16
109	62	59	11/16	36	361	1·1/8	1·1/8	1·1/8
115	62	59	11/16	36	359	1·1/8	1·1/8	1·1/8
121	62	59	11/16	36	346	1·1/8	1·1/8	1·1/16
127	62	59	11/16	36	323	7/8	13/16	13/16
133	62	59	11/16	42	352	7/8	7/8	15/16
139	62	59	11/16	42	371	15/16	1	1
145	62	59	11/16	36	377	1·3/16	1·3/16	1·3/16
151	62	59	11/16	36	370	1·3/16	1·3/16	1·1/8
157	62	59	11/16	55	355	15/16	13/16	1·1/8

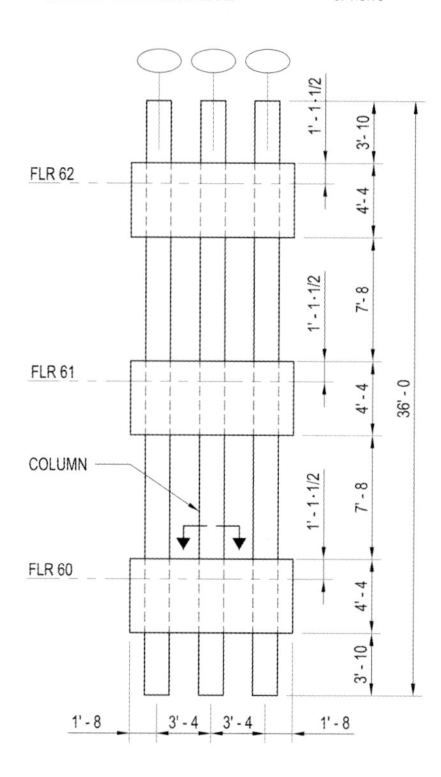

PREFABRICATION OF STRUCTURAL STEEL

We had used extensive prefabrication of structural steel for other projects, most notably the IBM Building in Pittsburgh, but for the World Trade Center, prefabrication was attained to an unprecedented degree. Two examples will provide something of the scope:

Steel plates were produced and fabricated in Japan before shipping to Washington State for assembly by Pacific Car and Foundry. The prefabricated panels were 3 stories high and 3 columns wide (36-feet high by 10-feet wide), and shipped by rail across the country to the site. Similarly, 60-foot-long and 10-foot-wide floor panels, complete with profiled metal deck and electrical distribution cells, were assembled in New Jersey from components fabricated in Missouri and elsewhere.

The process was nearly seamless. Why? We had taken all parts of the World Trade Center, almost every piece of steel, and described them in a digital format using IBM punch cards. For the 3-story wall panels, the size and grade of each of the plates, every weld, every bolt, all dimensions, were provided in these punch cards.

COMPUTERIZED PURCHASING AND FABRICATION

This was probably the first time that the structural designs for a high-rise building were provided to contractors in a digital format. A computerized system was conceived and developed for ordering structural steel, for producing the structural steel shop drawings, as well as for the operation of computer-operated tools, all directly from digital information that we developed as a part of our designs.

Traditionally, drawings had been the legal basis for work to be accomplished under construction contract. It is interesting, at least to me, that our specifications, which called for the digital data to be used as the basis of the design, were negated by Port Authority's legal staff without first consulting with us. Ultimately, reason carried the day. The many fabricators (as I recall, there were 24 principal fabricators) struck back, demanding that our IBM punch cards, containing the digital data, should be the basis of the contract and asked for additional money should this not be the case. The legal staff backed down but not without considerable study.

The digital process was accomplished on an IBM 1620 computer, the most powerful medium-scale solid-state computer of its time. Each of our external drives was the size of a washing machine. The multiplexer was the size of a refrigerator. We used IBM punch cards for input onto the disks; output was by line printer, the size of a desk but taller, and by the IBM punch cards. IBM's name for the machine was the "Cadet." Once realizing that the Cadet's addition was by looking up numbers in tables and that multiplication was achieved by multiple addition, we understood that the name meant Can't Add, Didn't Even Try.

Right Engineers at our IBM 1620, IBM's first medium-scale, solid-state (without vacuum tubes) computer.

Bottom right The first of its kind, a wind tunnel model that included proximity models of the surrounding cityscape. Since the towers were identical in their pressure-loading systems, only one pressure model was used and then interchanged with the empty dummy model.

Below A separate model reveals pressure transducers inside one of the low building models.

WIND ENGINEERING

At the outset of the design for the World Trade Center, my knowledge of wind engineering was minimal, more or less in keeping with that of the majority of the structural engineers of that era. It was Malcolm P. Levy of Port Authority who came to us suggesting that there was to be the first Wind Engineering Congress, and that it would be held at Teddington, UK. It was my first experience outside of the United States and my first experience at a structural engineering congress. Following a few days at the congress, it became clear to me that Professor Alan G. Davenport's mind was miles ahead of all others at the congress. Further, following repeated discussions with him, I came to understand that Alan would be an essential component in our design team and that it was imperative that we convince him that we had a great need and that he had much to gain in joining us. With his family in Canada and his professorship at the University of Western Ontario, there would be the need for him to commute, weekly, between London, Ontario and New York. Alan, understanding perhaps better than I that this would be a not-to-be-missed opportunity for the furthering of the science of wind engineering, agreed to join our team.

Realizing that the World Trade Center was to be constructed on one of the windiest sites in Manhattan, we mounted a comprehensive program to determine the design-level gradient wind speed for New York City (the wind speed high enough above the surface of the earth so as not to be significantly slowed by surface roughness). Data was collected from all available sources, including the US Weather Service, with records dating back to 1873, and incorporated into appropriate mathematical models. For the first time, we were able to obtain full-scale measurements of the turbulent structure of the wind. This was done by mounting anemometers atop the Telephone Company Building immediately adjacent to the site, as well as atop two other high points in Lower Manhattan and then projecting the data to the site of the proposed towers.

Fortuitously, Martin Jensen and Niels Franck, two brilliant Danish engineers, had come up with the idea of the boundary layer wind tunnel, which they had used to accurately predict wind pressures and suctions on small farm buildings. Borrowing this concept, and with enormous input from Professor Davenport, who worked in our office for more than two years, we expanded this technology upward to 110 stories. We used a wind tunnel, constructed under the guidance of the late Dr. Jack E. Cermak of Colorado State University, originally designed to study the dispersion of gases emitted from tall smoke stacks. This was the first time that a building had been analyzed for the steady-state and dynamic components of wind-induced deflections on the structure and on the glazing, all under the influence of a turbulent (realistic) wind. This is now standard procedure in the design of tall buildings worldwide.

Using our new technology, with the most of the architectural, structural, and mechanical, electrical, and plumbing plans essentially completed, we discovered that, by rotating the services core of the South Tower by 90 degrees, the resulting change in gravity loads on the perimeter columns produced a useful improvement in the wind-induced dynamic response of the towers. After checking and rechecking this finding, I took a flight to Detroit, there to explain our findings to Yama. His first response, "This is not possible," was tempered by Aaron Schreier, who delayed further discussion until after lunch, by which time drawings could be provided to show the changed relationships of the two giant lobbies. Yama agreed to the change that afternoon, conditional on acceptance by Mal Levy, which was readily achieved.

A more serious situation arose when we found that—through eliminating one of the columns on each face of the towers, along with a small increase in the width of each of the columns of the perimeter wall—we could achieve a significant reduction in the dynamic responses of the two towers, resulting in important cost savings. Inexplicably, Yama and Port Authority accepted this change without major comment.

This being in the pre-CADD era, all of the drawings had been accomplished by those wonderfully talented folks by title "drafter" or "draughtsman." To change our drawings to accept the revised perimeter column layouts, our Wayne Brewer came up with the idea of accomplishing these changes photographically; Emery Roth & Sons redrew each and every one of their 250 plan drawings.

As well, we were able to reliably predict wind speeds in the surrounding streets and in the central plaza. Particularly for the plaza, I showed Yama a variety of remedial measures that would make the area more habitable. Yama, thinking of the plaza as a great Italian piazza, refused to consider my requests. After all, he reasoned, that which would work in Rome would surely work in New York. We all make mistakes.

For the street-level west entrance of the North Tower, subsequent to occupancy, we designed a canopy to protect visitors, arriving by automobile, from falling ice and sweeping winds; the plaza was not usable in winter.

MOTION SIMULATION

From our wind studies, augmented by our work in the wind tunnel, we knew how much the buildings would oscillate, but we didn't know how much lateral oscillation was acceptable to the buildings' occupants. Seeking definitive criteria, I first met with organizations that had conducted motion experiments. Several allowed me to ride in their equipment, which gave me a profound appreciation of the ability of fighter-pilots, astronauts, and their ilk to endure the stomach-wrenching rides in such simulators. Clearly, beyond a kind of academic interest, there was nothing to be learned from these prior experiments.

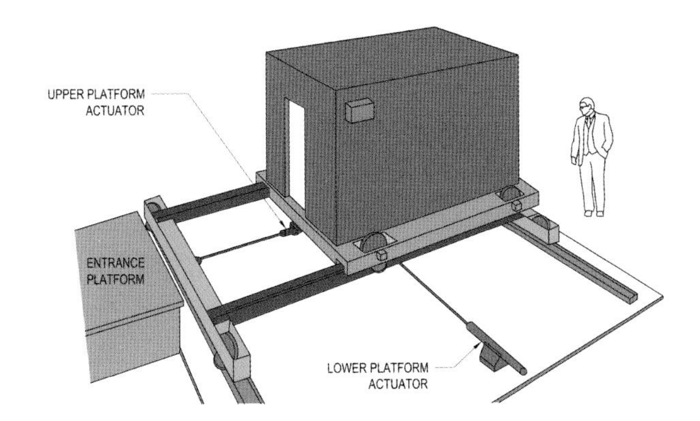

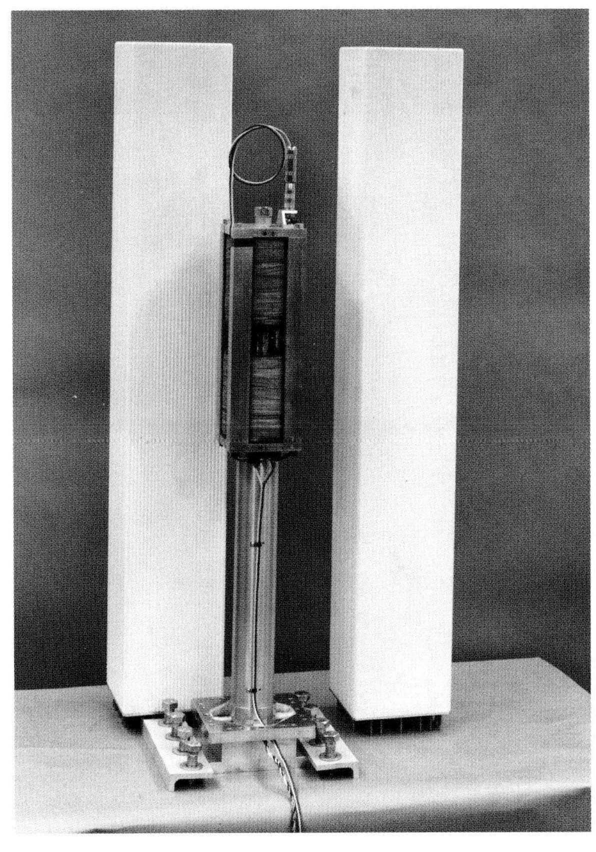

Simultaneously I had turned to more down-to-earth resources, seeking advice and guidance from specialists in human psychology. All opined that people were accustomed to lateral motion, that they would accommodate; there was, they said, no need to worry. Still, this was a billion dollar project (roughly $6.2 billion in 2016 dollars) to be built at an unprecedented height with no comparative studies available. We could not simply accept such advice unchallenged. Accordingly, as far away from the New York press as possible, we designed and had constructed at the Oregon Research Institute a motion simulator that more or less replicated the lateral swaying motion of our towers and measured the responses of persons in it. We learned right away that this motion was not immediately perceived by most subjects, but once perceived, was disturbing. That is, we had a perception problem, not a tolerance problem. Yama was fascinated with our work, taking the time to travel to Oregon to experience the sensation of the oscillation as would later take place.

Top left A diagram of the motion simulator at human scale.

Above left Yamasaki seated in the motion simulator.

Above Pioneering aeroelastic models were developed to measure the aerodynamic response of the individual buildings in turbulent wind. The motion-sensing transducer is seen removed from the outer tower shell at left.

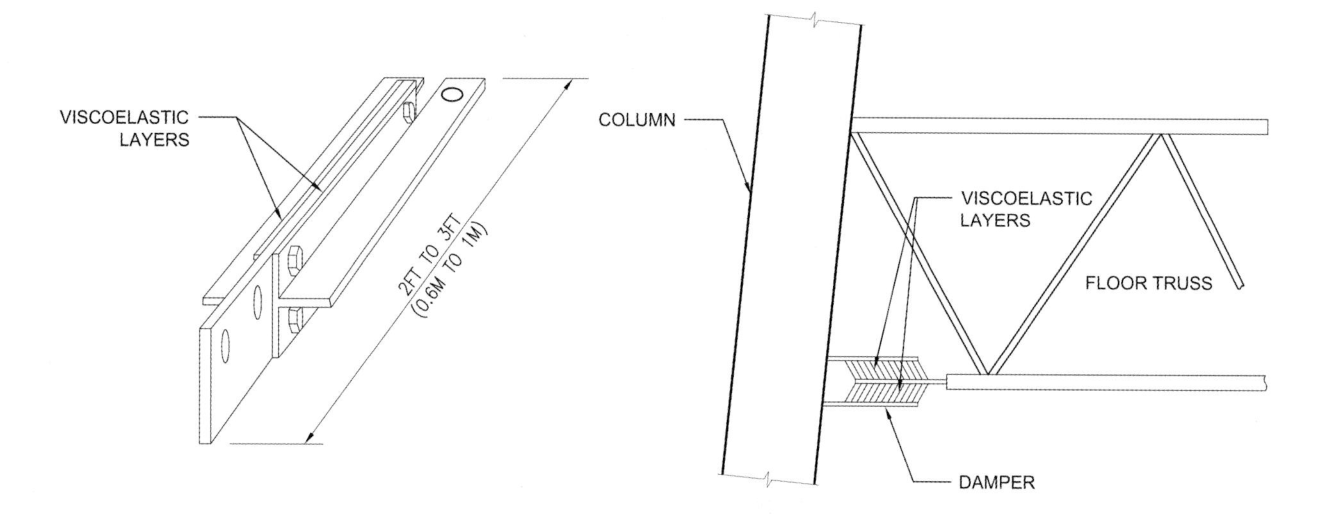

Above and above right Isometric of a typical damping unit and illustration of the interaction between the damper, perimeter column, and a floor truss.

Center The connection between an installed damping unit and its interacting column and floor truss.

Bottom Ray Monte of Port Authority examines the downward deflection of perimeter glass atop a vacuum box. Groundbreaking in the 1960s, glass-testing is now an industry standard.

To allow its executives to experience the building motion, Port Authority elected to construct a second simulator,to be hung as a pendulum. But there was no aircraft hangar or other suitable building tall enough to create a pendulum with a natural period of oscillation of approximately ten seconds, which was the first-mode natural frequency of the structures of the World Trade Center. Not to be discouraged, Port Authority found an airshaft located in a cross-Hudson River tunnel that fit the bill. The simulator was constructed, fortunately with findings not incompatible to our own. The data from our simulator was used to establish the criteria for an acceptable level of swaying motion of the two towers.

STRUCTURAL DAMPING

While our various studies provided us with essential information, the need to limit oscillation to acceptable levels fell squarely under our responsibilities. To ameliorate the wind-induced dynamic component of building motion by dissipating much of the energy of that motion, under the watchful eye of Prof. Davenport, we invented a viscoelastic damping system and patented the system with the cooperation of the 3M Company, fabricators of the dampers. Acting more or less like shock absorbers in an automobile, the damping units formed a structural system perpendicular to the primary system, allowing us to control the swaying motion without having to use large quantities of structural steel. This is the first time that engineered dampers were used to resist the wind-induced swaying motion of a building. Yama was intrigued by this concept but somehow took it as just the sort of thing that we engineers are expected to do. The executives of 3M, worried that the dampers were critical to the safety of the building and the potential liability this involved, were less enthusiastic; it took my meeting with the highest level executives of the 3M Company to convince them that the dampers were for human comfort, not for human or building safety.

DESIGN OF THE GLASS

The design of the aluminum and glass facade remained a knotty problem as, again, there was no comparable design procedure. Realizing that the breaking strength of the glass was not tested as a component of the manufacturing process, Dr. Peter Chen of our company and I developed a theory for integrating the statistical strength of glass with the dynamic (fluctuating) forces of the turbulent wind found in the wind tunnel testing. Our theory allowed a prediction of the breakage rate of the exterior wall's glass. For the first time, coupled with a testing program of full-scale glass samples, we were able to rationally determine the necessary thickness and grade of the glass, arriving at the selection of semi-tempered glass in 19-inch-wide by 78-inch-high panels.

Of interest, the glass did break but not from the wind. Some person lobbed bullets at the towers, sometimes striking glass and sometimes the aluminum cladding; of course, the bullet impact broke the glass. Another individual, driving on the West Side Highway, propelled 1-inch-diameter steel balls at the lobby glass. To the best of my knowledge, beyond a modest chip in the glass, there was no further damage.

STACK ACTION

We developed a theory to predict the stack action (akin to temperature-induced air flow within a chimney) and both temperature- and wind-induced stack action airflow within a high-rise building. An understanding of these airflows is crucial for the control of fire-generated smoke dispersion and for the reduction of the energy consumption of the building. Attempting to sift through the myriad of circumstances that could harm the structures or the occupants of the towers, fire and smoke were the most common issues. The technical paper that I had developed and published on the theory of stack action within a tall building provided a solid foundation for the amelioration of this problem. Stack (or chimney) action arises because of the differences in the weight of the air that is within the building and that which is without. The neighboring Pan Am Building provided an instructive case study of stack action. The entire lobby was open to Grand Central Terminal so that, in the winter, the rush of air from the terminal into the building created difficulties in the opening and closing of elevator doors and the like. At the World Trade Center, we were to build 70 percent higher, thus compounding the problem by the ratio of approximately 2:1. Proper airlocks around the towers provided the required control systems. We believe that the World Trade Center was the first building to rationally design for these air flows and pressures.

ELEVATORS

To minimize the oscillation of the elevator cables, which is stimulated by the wind-induced swaying motion of the building, I developed methodologies to predict appropriate "parking floors" for the elevators. The approach was to park the elevators such that the natural frequencies of vibration of the cables was significantly different from the natural frequencies of the towers. A comparison of the wind-induced dynamic components of the structural response of the two towers in juxtaposition to that of the Empire State Building is shown on the top of the opposite page.

AIRCRAFT IMPACT

Knowing that the Empire State Building had been struck by a Mitchell bomber in 1945, we designed the World Trade Center for possible impact from a Boeing 707,

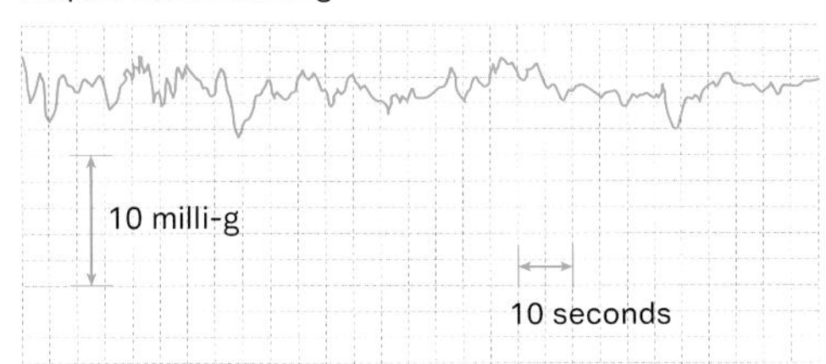

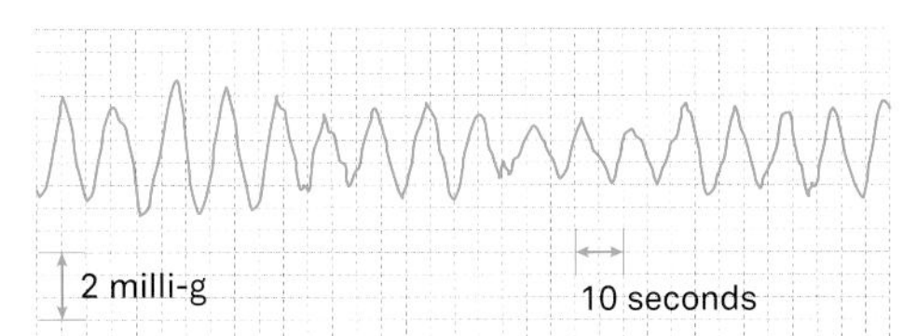

Top Graphs marking the significant differences in wind-induced dynamic lateral response in the Empire State Building and the World Trade Center. Evidence of these differences helped authenticate the design philosophies developed for the two towers.

a 4-engine jetliner that dominated commercial air transport in the 1960s. The two towers were the first structures outside of the military and nuclear industries designed to resist the impact of jet aircraft. It was assumed that the jetliner would be lost in the fog, seeking to land at JFK or at Newark Airport, i.e., a low-flying, slow-flying jetliner. To the best of our knowledge, little was known about the effects of building fires from such aircraft, and the architects did not prepare designs for that circumstance. Indeed, while the matter was outside of our responsibilities, at that time fireproofing systems were not available to control the effects of such fires.

PARKING UNDER THE TOWERS

This being the 1960s, my concern was not related to terrorist attacks, but rather with the spread of smoke from automobile or truck fires, which are not uncommon. Coupled with stack action, such fires would spread smoke and soot throughout the building. Parking under the towers, in my opinion, was dangerous and unacceptable. Of course, I took my concerns to Yama.

Having never heard of such things, Yama was more than a little skeptical, properly arguing that the loss of parking spaces would create a completely unacceptable financial burden. I pleaded with him, nearly going to my knees. "All right," he said, "we will take your concerns to Austin Tobin," Port Authority's Executive Director, the only person with authority to make the decision.

It was some weeks before Mr. Tobin and Yama were in New York at the same time and with clear calendars. When the meeting finally took place, Tobin listened thoughtfully and carefully to my concerns before turning to John M. (Jack) Kyle, Port Authority Chief Engineer. "What do you think?" he asked. Jack took a deep breath before responding: "This is something that we should

Below The North Tower under construction in late 1968, with the PATH rail line crossing through the site (bottom center). The South Tower and the basement framing were not yet underway.

have considered a long time ago, and it's strange that the concept should come from our Structural Engineer. Even so, Les is absolutely right. While I will clear it with our staff, for now, we should put together the consequences of removing the parking from under the towers." I was scarcely able to withhold my relief. About a week later, the parking under the towers was removed.

From this experience, the value of all design professionals becoming familiar with the larger picture becomes abundantly apparent.

METAL STUD AND WALL-BOARD PARTITION SYSTEM

In past decades, architects and engineers had not been concerned with the oscillation of their buildings, because the masonry block and brick partitions then in use added significantly to both the lateral stiffness and the structural dampening. As part of our investigation of tall New York buildings, we performed a computer analysis of the structural steel frame of the Empire State Building. By measuring the natural periods of oscillation of the building, we were able to measure the stiffness of its structural steel frame, comparing our field measurements with our computer analysis. As constructed, the Empire State Building was much stiffer than we had digitally determined. We carefully checked our work and measured again. The difference was shown to come from the brick and stone partitions in the building, which provided compression diagonals within the steel frame. The stiffness of the Empire State Building was due in one part from the steel frame and about five parts from the effects of the masonry in-fill. Engineers did not previously have the tools for determining the effects of such loading but learned from experience that the system performed well.

Early in the design I had been able to ride atop elevators so as to examine the effectiveness of gypsum block, fire-rated partitions used up to that time. From inside of the shaft, I could see light from the occupied spaces, indicating that air could move easily between the elevator shaft and the elevator lobbies so that, in the case of a fire, this air flow would carry smoke to all of the floors above the fire. In response, we developed a new kind of shaft-wall partition system, now widely used in place of gypsum-block masonry and plaster walls. At that time, masonry walls were the standard fire-resistive enclosure for elevators, stairs, access-ways, and other internal structures. Our new partition system, developed jointly by our Wayne Brewer and me, eliminated the need for within-the-shaft scaffolding, which was then common practice, while also reducing air flow due to stack action, providing more smoke-proof stairs and shafts and improving the safety of the jobsite. Shaft-Wall, or its equivalent, by eliminating the stiffness and the damping intrinsic in the masonry partition systems, completely changed the nature of the structural system for the two towers, making them the first of a new kind of high-rise building.

CONSTRUCTION SAFETY

The prefabricated floor decks and closely spaced perimeter column assemblies (each column being approximately 14 inches square, spaced 3-feet 4-inches on center) increased economies of construction time and cost while improving the safety of workers.

In large measure because of the construction safety regulations and inspections imposed by Rino M. (Ray) Monti of Port Authority, there were only eight deaths in the construction of the World Trade Center; none of these deaths were iron workers. In the past, for most projects, the majority of deaths were the result of a construction worker falling down a shaft. Beyond Port Authority, other contributions to this safety record came from our designs: the closely spaced columns forming a wall at the perimeter of the building, the prefabrication of the floor panels, complete with profiled metal deck, and the use of our shaft-wall partition system; the latter allowing quick construction of the walls around elevator, stairs, and other shafts.

PORT AUTHORITY

Working with the Port of New York Authority was, for me, a real pleasure. Being relatively inexperienced, I assumed that all these large organizations were similar: engineering-oriented, strong, and filled with talented men and women, ready to look to the future. Through all of our needs for research, Port Authority examined our proposals, sometimes proffered constructive suggestions, but always supported our efforts.

CRITICAL RESPONSE

In February 1964, shortly after the twin tower scheme was made public, Ada Louise Huxtable, the renowned architectural critic in one of her first articles in *The New York Times*, wrote: The World Trade Center "may herald a new era of skyscraper design for the city that made the big building its trademark." She continued: "From the design aspect this is not only the biggest but the best new building project that New York has seen in a long time. It represents a level of taste and thought that has been distressingly rare in the city's mass of non-descript postwar commercial construction."

In summary, she wrote, "The Trade Center faces the problems of big city building head-on, and does so with a civic conscience and an imaginative search for new and better solutions. The huge towers are planned as the stupendous focus at the end of a handsome plaza surrounded by lower, arcaded buildings conceived on a sensitively human scale."

Within a couple of years Ms. Huxtable began to have a change of heart and by the time the World Trade Center was dedicated in 1973, she deplored "The towers [as] pure technology [and] the lobbies [as] pure schmaltz . . . These are big buildings but they are not great architecture. The grill-like metal facade stripes are curiously without scale. They taper into the more widely spaced columns of 'Gothic trees' at the lower stories, a detail that does not so much express the structure as tart it up. The Port Authority has built the ultimate Disneyland fairytale blockbuster. It is General Motors Gothic." ("Big but Not so Bold," *The New York Times,* April 5, 1973).

Yama responded with equal eloquence and quiet gentleness in a thoughtful personal letter to Ada Louise. A copy of this perhaps previously unpublished communication is included here in its entirely to offer a singular and unusual view of Yamasaki, the man and the architect, and his design goals for the World Trade Center, but also equally to share a rare insight into the dynamic of architect and critic such that the words of each may be filtered through the lens of time. Years later, long after Yama had succumbed to cancer, I wrote to Ms. Huxtable to request her willingness to make the letter public. I never did hear back but include here as a matter of historical record the copy that Yama presented to me, and which I helped, in part, to draft.

Returning to the excitement that greeted the release of the World Trade Center design in 1964, Wolf von Eckardt of the *Washington Post* declared the project "a magnificent work of architecture and urban design." He went on to say that "The [towers'] skillful and handsome design and the magnificent open plaza on which they are to rise can evoke only unqualified admiration."

But like Ada Louise Huxtable, nine years later, von Eckardt, wrote:

"There is a fascination in the towers' ugliness . . . Man's tallest buildings to date defy their surroundings, man's most wondrous skyscraper community. . . . The 110-story Broddingnagian shafts stand with blunt, graceless arrogance at the western edge of Manhattan island, seeming to tilt that wonder with overbearing size and hubris."

Other critics joined the deprecating chorus. It is easy to respond to such pendulous sways with either anger or humor. While I prefer the latter, and in full realization of the changes that were taking place in the dominant style practiced by the architectural profession, it is easy for me, as one who came to love Yama as a dear friend, to carry some resentment toward the practice of the architectural critic.

THE POST-CONSTRUCTION PRESS

When completed, the towers were opaque from most vistas, but when the sun set behind the buildings their airy lightness could be appreciated. They continued to be admired by sculptors, but Yama's elegant designs for the World Trade Center fell completely out of grace with the architectural community and, indeed, with New Yorkers in general, who found the towers cold and too large. It wasn't until 1974, a year after the project opened, that the indomitable Philippe Petit strung his wire between the the towers, performing thereon for well over an hour, and in the process, leading New Yorkers to embrace the towers and recognize them as proud icons of the city.

A year after Philippe's dramatic excursion, Owen J. Quinn parachuted from the North Tower, followed two years later by George Willig's climb up the outside face of the North Tower. None of these feats contained the artistry of Philippe's walks but Quinn, particularly, received considerable press over what seemed to me, having ice and rock climbing experience, to be a purely technical undertaking.

Twenty years were to pass, indeed it was long after Yamasaki's death, before the architectural community again began to view Yama as one of the most significant architects of his day. Much to my astonishment, I was invited to the School of Architecture at Princeton University, as critic for designs prepared by students for the World Trade Center. It was gratifying for me to see and to feel the admiration of the students for this amazing man, this most complex architect, this dear friend, Minoru Yamasaki.

Shortly after dawn, one of the countless examples of the sun's always dynamic relationship with the towers.

Minoru Yamasaki
350 West Big Beaver Road
TROY, MICHIGAN 48084
April 10, 1973

Mrs. Ada Louise Huxtable
The New York Times
969 Park
New York, New York 10028

Dear Ada Louise:

Since you became the architectural critic for the Times, I have been very happy, because I knew that finally there was a sensitive critic with the knowledge of present-day construction to relate to the public what we architects are trying to do. Normally I pay little attention to criticisms written about our buildings; sometimes they are fair, sometimes they are prejudiced and sometimes they lack an understanding of our designs. Therefore, I was surprised to see your comments on the Trade Center last week.

We have been working on the project for ten and a half years, and throughout this period we have made many design decisions; as they have been resolved into actual construction I have felt that they have resulted in the kind of buildings which are as I had hoped they would be.

In the middle of the last century Emerson stated principles which might well form the basis of a philosophy of architecture. He said: "Beauty rests on necessities. The line of beauty is the result of perfect economy. The cell of the beehive is built at that angle which gives the most strength with the least wax. The bone or the quill of the bird gives the most alar strength with the least weight."

"There is not a particle to spare in natural structures. There is a compelling reason in the uses of the plant for every novelty of color or form; and our art saves material by more skillful arrangement; and reaches beauty by taking every superfluous ounce that can be spared from a wall and keeping its strength in the poetry of columns."

He continued,

"If a man can build a plain cottage with such symmetry as to make all the fine palaces look cheap and vulgar; can take such advantages of nature that all her powers serve him, making use of geometry instead of expense, tapping a mountain for his water jet, causing the sun and moon to seem only the decorations of his estate, this is still the legitimate dominion of beauty."

My work has been affected very much since I read this passage many years ago.

Rather than having enormous exterior and interior columns which interfere with the activities of people on the many floors, we chose to make our exterior wall both load-bearing and the cantilever truss which stiffens the building against the very high winds found at the upper levels, the most difficult structural problem in a high-rise building. The cores in the Trade Center towers contribute nothing to the stiffness of the buildings.

As far as "lace" is concerned, it is the valid component of the truss. We have three columns 40" on center above, which translate to a single column at the base of the building, giving what you call a sense of Gothic design. In my view, the concept of the wall, its relationship to the mass of the building and its details are consistent with Emerson's thoughts: "it reaches beauty by taking every superfluous ounce that can be spared...keeping its strength in the poetry of the columns." I certainly have no shame in this, because the wall is a purely and thoroughly economic way to bring the load down and makes possible sufficient openings at the base for people and materials to easily pass through. This delicate wall, which you call "dainty," is not only a very beautiful truss, but it carries floors with spans of sixty and thirty-seven feet. These floors are completely column-free, giving maximum freedom in circulation and in office and furniture arrangement. To me, this is one of the gifts of our technology, when we can build 110-story buildings with both human-scale elements and the advantage of long spans.

For me, the day of the all-glass building is finished. The problems which come from lifting and installing enormous panes of glass, which are then shaded by curtains more than half of the day, are almost ridiculous. When glass size is pushed unreasonably, there are structural problems which arise; as you are aware, on several recent occasions in two of our major cities, large panes of glass have been blown out, to the point where buildings had to be barricaded and the pedestrians protected. Large glass, as lovely as it may be, requires a tremendous consumption of energy for heating and air-conditioning to combat the extremes of temperature. Moreover, glassy buildings tend to be "curtain displays"--some closed, some half-open, some open--which adds a more restless quality to an already restless city.

I certainly agree that there should be large glass in observation areas where the prime purpose is just that, to give a panorama of spectacularly beautiful views. In normal working conditions a reasonable amount of glass is of course necessary, so that people may be aware of whether it is beautiful and sunny, or rainy and miserable outdoors, giving them contact with the outside world. It also should be present in sufficient quantities to give relief to the frequently monotonous work which goes on in the normal office. In many all-glass buildings the people who work and live there have a strong sense of acrophobia, which makes their lives uncomfortable.

I must ask myself if we want to design buildings for people to fit some preconceived idea of a glass world. Is this really the future of cities? As for mirror glass, I detest it, because buildings with it look to me as if they have cataracts, showing no life within. On the interior, it produces strange reflections of lights, objects and people which gives me a feeling I can only describe as eerie. As regards the narrow windows, they give me none, if any, sense of acrophobia; and since they are narrow, they cannot bond to the point of blowing out under gale conditions.

Recently The Times carried an article about many lower buildings which moved a great deal in sixty to seventy mile winds so that their upper floors had to be evacuated. These are buildings with much shorter spans, designed for only a little more than one-half the wind force for which the Trade Center buildings are designed. As a matter of fact, the New York Building Code has revised its wind criteria upwards since we designed our buildings, and our standards are still higher. Our buildings are so designed that on the two-to-three-decade cycle when winds reach hurricane force, there may be some discomfort, but there will be no oscillation, as experienced in the buildings described in your newspaper.

As you know, the land in New York is very expensive; hence, the almost impossible overloading of the Wall Street area with its narrow sidewalks and its great confusion of traffic. Often I feel when I walk there, that I am about to be pushed into the street, to be hit by an automobile. The big building can resolve this, as we have done in the Trade Center, where we have a five-acre plaza where people can enjoy their moments of leisure, the sun and the plants. If you will look closely at the Trade Center, the sidewalks on its perimeter are very wide. They go in and out, without the usual parallel relationship of building to curb, so that people can walk in an uncrowded and comfortable way.

For the lobbies, I cannot understand why you called them "schmaltzy." To me, "schmaltz" has always implied buildings like the former Huntington Hartford Museum. Our lobbies have no applied details--just fine materials, and the only elements which may be decorative are the crystalline lights seventy feet up, which light the great spaces. The reflective surfaces interestingly reflect the people passing through the space, bringing them into the building in an unexpected and pleasant way.

In the limited number of projects on which we are privileged to work, our primary objective is to heighten the quality of experience for those who work or visit or live in the buildings.

Now, imagine the daily experience of the 20,000 or more people who work in each of the towers, who travel on the grim subways (or even on PATH). We hope that this fifteen-acre retail area will provide both fun and interest on their way to the sudden surprise of the tower lobbies--sun-filled spaces seventy feet high--before they take their elevators to their respective offices. That was planned; it did not just happen. Also, consider the fact that the shuttle/skylobby elevator system--55 passenger cars traveling at 1600' per minute, and which we had to plead for--transports the person easily to the upper floors. These cars stop only once; with doors at both ends, the first passenger on is the first one off; he then transfers, at the skylobbies, to a normal bank of elevators to his ultimate destination. This system eliminates a tremendous number of "dead" elevator shafts and makes a significant addition to the useful and rentable floor area of the tall buildings.

For the 500' x 1000' excavation west of Greenwich Street, the engineers had originally planned to use sheet piling to a depth of only thirty-five feet, using conventional piles for the tower foundations the additional thirty-five feet down to rock. The sheet piling would have meant that a large portion of the Atlantic Ocean would have been in the hole, requiring fantastic pumping during excavation. It took six months for us to convince them that there must be a better way, and the slurry wall which you see in the attached brochure is the result. Now we bear all main columns directly on the rock and we were able to build three floors, virtually for free and save millions of dollars by practically eliminating the dewatering. This area houses the new PATH station, refrigeration plant, service trucking, storage and garage areas. Material which came from the seventy-foot-deep excavation made twenty-six acres of usable land on the Hudson just south of the Trade Center. This I sincerely hope will be turned into housing and recreational areas which are so badly needed in this area of Manhattan.

You may also be interested to know that there are two million square feet of exterior curtain wall in the Trade Center towers and very little sealant; the little which we have is protected and need be maintained only at long intervals from the interior. This I made a rule very early in the design, because I just couldn't see a man hanging 1350 feet in the air re-sealing the joints of the building on a constant basis, and still be plagued with leaks; this is the major problem of maintaining the thin walls our technology has made possible. As a result, the exterior walls of the two towers have had no leaks.

With the use of the systems above it was possible to build the towers for about the same price as the Pan Am Building when compared in constant dollars, which is a fair achievement considering the relative heights of the buildings.

I am not implying by this letter that these buildings are great. That is perhaps not for either of us to say, but for the people to decide in the many years during which the buildings will live. I do think that the plan of the Trade Center allows outdoor space for people at ground level--happy, pleasant space undisturbed by automobiles.

I am happy I was able to design very large buildings which have the scale relationship to man so necessary to him; they are intended to give him a soaring and inspiring feeling, imparting pride and a sense of nobility in his environment. They are set back five-hundred feet from Church Street; their changing quality as one approaches across the plaza is, to me, greatly inspiring. So many tall buildings say nothing at all when you are next to them; or their great beams and columns are gloomy and fearsome from directly below, as they sit so solidly and so closely to the sidewalk and street.

Finally, these advances in building technology allow us to economically compete with lower buildings, giving opportunity to open up our urban centers with spaces, be they green or paved, which are so important to life in the heart of our cities.

I hope you take this letter in the spirit in which it is intended; I have no malice because of the article.

My best wishes to you.

Sincerely,

Minoru Yamasaki

Mr. Minoru Yamasaki
350 West Big Beaver Road
Troy, Michigan 48084

Dear Yama,

I am most appreciative of your long and thoughtful letter on the World Trade Center. I also appreciate the fact that after ten years of conscientious work, my reflections would not make you happy, and that makes me less than happy too, but you know it is my job to call it as I see it. I tried, in the brief space I had, to make clear that you had a philosophical and structural rationale for your design; but I feel the aesthetic effect is not really successful for buildings on this scale.

As you probably know, I did an article some years ago on the engineering of the buildings, so that most of what you wrote is familiar. As engineering, they are extremely impressive. I suppose that it is design approach, or interpretation of structure, that is the issue, and that, admittedly, can be a very subjective matter. You are the architect, and I am the critic, and it is an honest parting of the ways. You are quite right that the future alone will really judge the result.

I am pleased that you cared to write me in the detail and spirit that you did.

Sincerely,

Ada Louise
April 23, 1973

It became apparent that the World Trade Center project was far too demanding for any but the very largest fabricators of structural steel. After some discussions with Mal Levy of Port Authority we focused our sights on Bethlehem Steel Company (Bethlehem) and the American Bridge division of United States Steel Corporation (AmBridge), for whom we were the structural engineers for the company's headquarters in Pittsburgh.

For the larger components, most notably the column/spandrel assemblies that formed the prefabricated wall panels, we completed separate structural drawings for Bethlehem and for AmBridge, in this way hoping that we would obtain the very best price at the needed quality. Indeed, both Bethlehem Steel and AmBridge, at regular intervals, provided budget estimates, met with us and otherwise cooperated fully. We knew of no other project that had taken this approach; understandably, the Port Authority legal staff was skeptical, concerned that this bordered on illegal contracting policies. While some changes were made to the legal jargon in the technical specifications, two separate sets of bid documents were completed and issued, one to AmBridge and the other to Bethlehem.

The two buildings were architecturally almost identical but structurally unique, largely because of the differences in yield-point of the steels used for the plates of the perimeter wall panels, which differences resulted from the wind-induced response of each tower. I was told that the contract drawings weighed more than 600 pounds! The two towers consumed about 185,000 tons of structural steel or about 30 pounds of steel for every square foot of floor space in the towers.

On August 17, 1966, when the bids were opened, the quoted prices were about 50 percent higher than those in the budget estimates. Of interest, the prices of the two steel companies were within three or four percent of each other.

With the prices so high, there was serious discussion of abandoning the project or of changing it drastically. To the best of my knowledge, it was John Tishman, of Tishman Realty and Construction Company, who proposed that the structural steel contracts be broken into small packages, catering to smaller steel fabricators world-wide. This strategy, ultimately saving some 30 percent over the AmBridge/Bethlehem bids, was accepted. For us, it meant that we were required to coordinate the shop drawings created by a myriad of fabricators, a process made possible because of the punch-card details provided by us.

Shortly after the hair-raising opening of bids, my phone rang with a call from the US Justice Department. The caller explained that they were investigating possible antitrust wrongdoings on the part of AmBridge and Bethlehem. I stopped him, asking if there were any restrictions on my making use of the information gleaned from his call; he assured me that there were no such restrictions. I then answered his questions regarding our relations with AmBridge and Bethlehem. Along the way I discovered that all of Port Authority's principal players in the bidding process, but not AmBridge and Bethlehem, had been interviewed.

Structural engineer Guy Nordenson's architectural students at Princeton University prepared this masterful drawing based on our data. Visualizing the varying yield points of steel for the perimeter columns of each face of the two towers, red indicates the highest-strength steels; blue is the lowest. Though the two towers were visually identical, in terms of engineering they were quite different, individually responding to the varying direction, speed, and turbulence of the wind on each of the four faces.

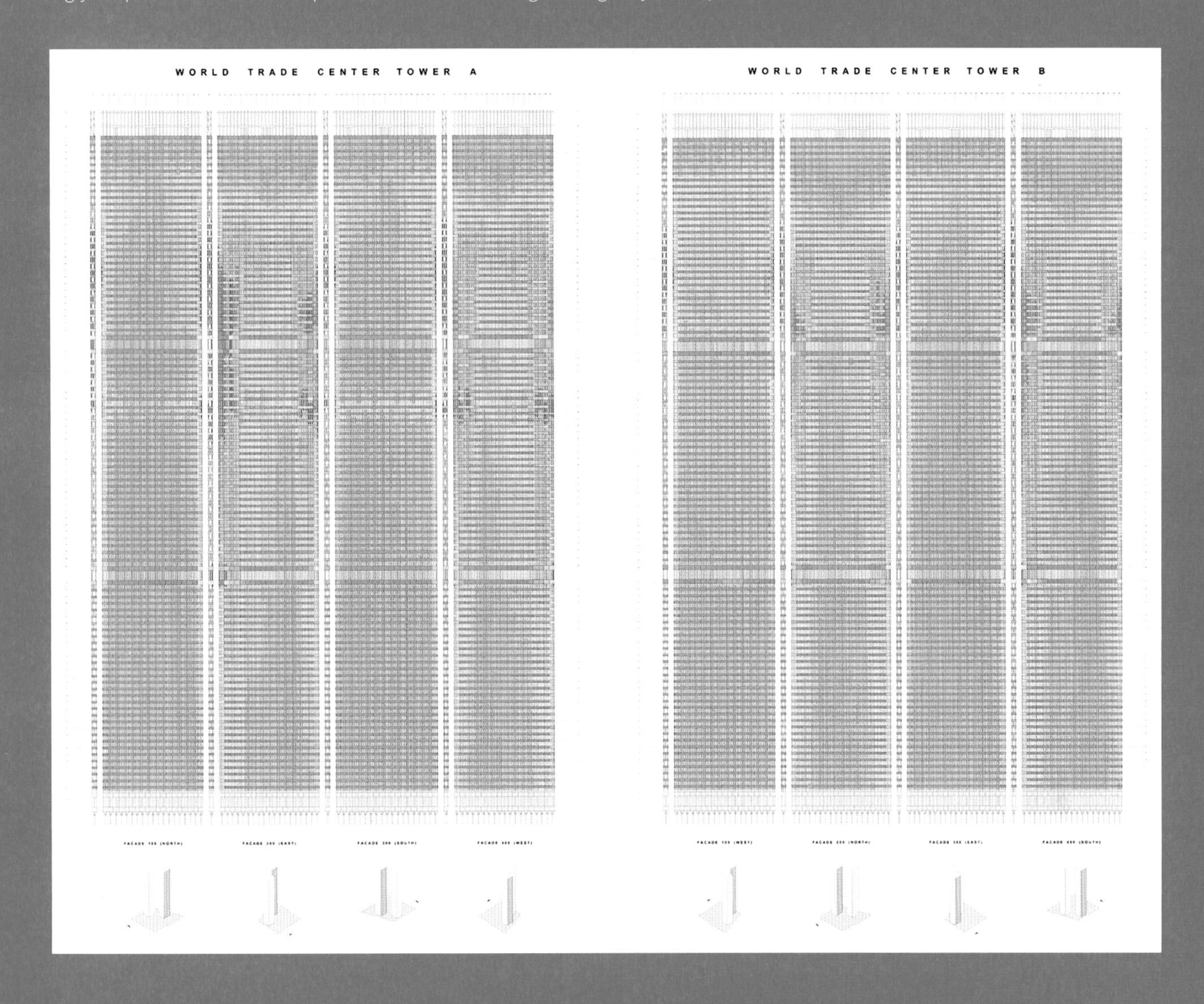

When the call was completed, I closed my door as I pondered what had been said. Then, separately, I called the presidents of AmBridge and Bethlehem. To each I recounted the content of my telephone conversation with the attorney from the Justice Department; both expressed astonishment that there was an investigation underway. Later, each, in turn, called back to thank me for keeping them informed.

All of this was worrisome to me. In calling AmBridge and Bethlehem, had I done something wrong? Should I have first consulted with an attorney? Of importance, why had not others—Tishman, Port Authority, Yamasaki, Roth—made the same phone call? After much soul-searching, I decided that my instincts were correct, that being frank and open with those with whom I dealt professionally was the proper path.

MALCOLM P. LEVY

Malcom P. Levy, who oversaw the design and planning effort for Port of New York Authority, was hard-working and brilliant. He was strong, he was a New Yorker, he was opinionated, he was knowledgeable, he was simply wonderful. While he was known for shouting his frustrations at those who somehow did not meet his expectations, only once did he raise his voice to me—a circumstance that we later sorted out in private. He and I, operating on completely different planes, were not "friends" in the sense of companionship, but we were compatriots. Mal Levy and Ray Monti, also of Port Authority, were the key figures in making possible the planning, design, and construction of the World Trade Center.

Many years into the project, I received a letter from a Port Authority lawyer asking if I had lunched with Mal on a specific day. As it turned out, on that day I was in Venezuela, and responded accordingly. Not long after, I received a letter from Port Authority's legal department stating that Mal was being brought up on charges of having falsified his expense accounts. I immediately sent a letter to Executive Director Austin Tobin pointing out that Mal was an extremely dedicated executive; that Florence (his secretary) almost surely made up Mal's expense records from such confused notes as Mal would provide, which he must have signed without a reading; that his service to Port Authority was exemplary; and that the entire charge was preposterous. Further, I wrote in praise of Florence, who, faced with Mal's impetuous nature and her need to file such reports on time, did the best that she could, which was not inconsiderable.

Later, I was called to a hearing to state my case in the matter, which I did, and with force. I was astonished to find that none of the other consultants—architects, engineers, and their ilk—felt the need to testify on Mal's behalf. It is hard to understand such things. As was his nature, Mal didn't call or write to express his appreciation. As was my nature, I didn't call or write to the myriad of consultants who did not write or testify to support Mal Levy. Well, I did discuss the matter with Aaron Schreier, who shrugged his shoulders, and gave me a big smile, before changing the subject.

Paul Hoffman of the Oregon Research Institute, Malcolm Levy, Minoru Yamasaki, and Aaron Schreier of the Yamasaki office at the research laboratory for the motion simulator.

THE BOMB
26 FEBRUARY 1993

Past noon, we heard on the news that some generators at the World Trade Center were on fire. This being a matter for the Fire Department, SawTeen, our daughter Karla Mei (who was nine years old at the time), and I drove to our weekend home on Candlewood Lake, Connecticut when, in the afternoon, a contractor arrived to look at a small problem. He mentioned nonchalantly that there had been a huge explosion at the World Trade Center. We jumped into our car, leaving Karla Mei with friends before setting out for downtown New York. As we neared the site, we were stopped and stopped again. I had with me a very official-looking identification card from the National Academy of Sciences: Investigation Team, Committee on Natural Disasters, which, though not being relevant to the matters at hand, got us through the barricades and onto the site.

With the goal of bringing about the collapse of the two towers, perhaps with the North Tower knocking down the South Tower, the perpetrators executed a daring plan: a 1,300-pound homemade bomb carried in a rented truck was detonated in the public parking area of the B2-level basement, just six or so feet outside the south wall of the North Tower. Fortuitously, as previously discussed (page 125), we had been instrumental in preventing parking and truck access inside and beneath the two towers.

Six people were killed by the blast and more than a thousand injured from smoke inhalation and related causes. Because the bomb had destroyed the air lock (revolving doors and the like) that had surrounded each tower, and because it was a typical, cold New York winter, the smoke and soot from the blast had been pulled upward into the towers by stack action (as in a chimney), causing untold hardship to some 50,000 persons seeking to exit the building. By this action, the smoke cleared quickly from the blast area. Except for the fact that there were no lights, visibility in the basement was nearly normal, making our examination of the damaged areas more straightforward.

The devastation to the underground reinforced concrete structure outside of the tower was unbelievable to us, creating a 100-by-200-foot hole and collapsing the ceiling of the PATH transit station. Much of the damage was beneath the hotel at the southeast end of the site, which is to say, closest to the detonation. Fortunately, the columns under the hotel had survived, but the loss of concrete floors laterally supporting those columns placed the hotel in grave danger. Some 5,000 tons of concrete and wreckage lay in a pile of unknown stability.

On the night of the blast, SawTeen joined me for an exploration of the basement areas. We worked our way down to B5, the lowest level, wading knee-deep in

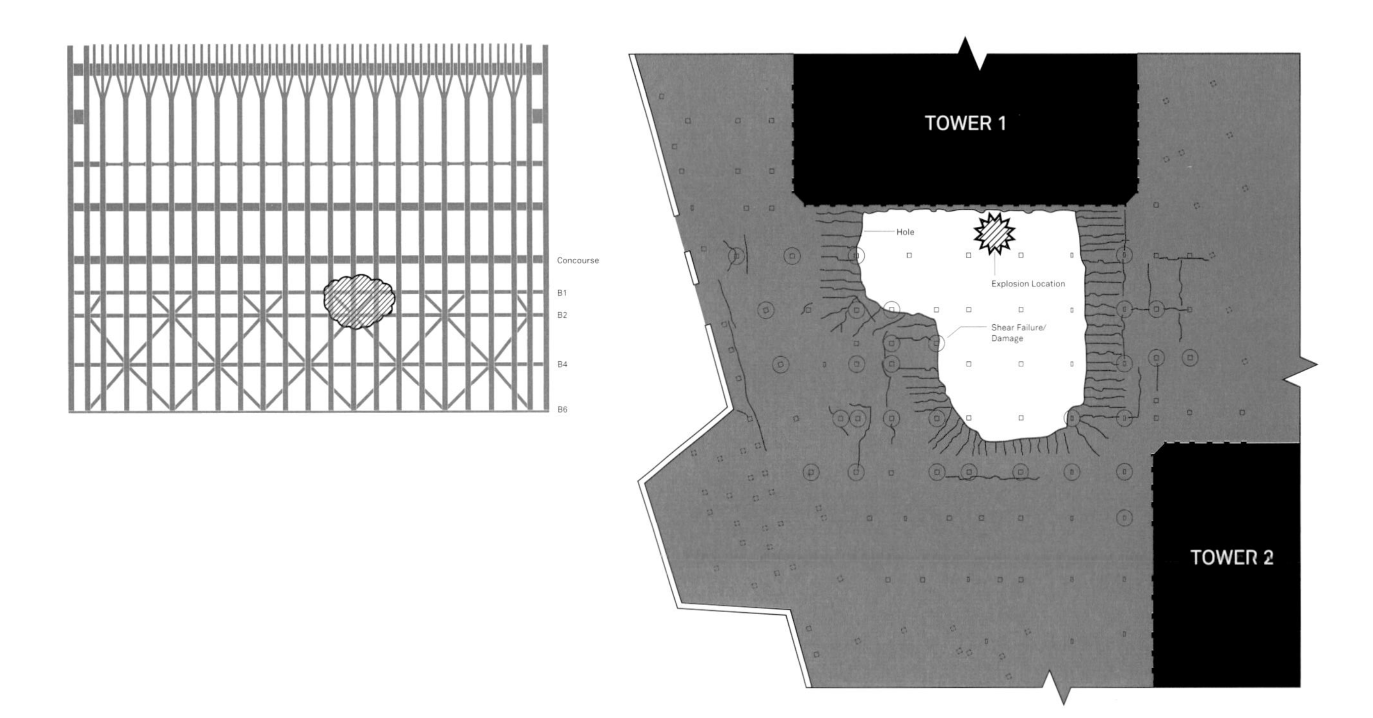

water, debris falling around us, the only light coming from our small flashlights. We returned to edge of the tower where SawTeen sketched some repair details. Indeed, before retiring on that long night, we completed sketches of the essential work required to stabilize the damaged columns supporting the hotel above, so that we effectively designed the concept for the bracing on the same day as the bomb went off. Fortunately, Karl Koch Erecting Company, which had erected the steel for the towers, had all of the required materials in hand.

Early the next morning, attempting to sort out the extent of the damage and now armed with a larger flashlight, I accomplished a solo tour of the lowest basement, discovering that much of the fallen concrete had landed atop the seven heavy (7-ton) air conditioning chillers, which were laterally supporting the perimeter concrete slurry wall, which was holding out the Hudson River and the waters of New York Harbor.

I returned upstairs to join a meeting of upper-level technical people from Port Authority, there to report my findings. We were alerted that Port Authority, the police, and others were meeting in one of conference rooms of the hotel . . . directly above the blast. Realizing the precarious nature of the supporting columns, Jim White and I pushed the panic button, asking immediately that the meeting, and all subsequent meetings, be elsewhere. Our advice was followed.

Clearly, the lateral support of the perimeter slurry wall was one of the most important issues before us. Port Authority officials also believed, quite rightly, that the restoration of the PATH transit system was at the top of the list. On Monday morn-

Top left The bomb was located in a parking area two levels underground, about six feet from the south face of Tower 1 (North Tower).

Above The area of total collapse of the reinforced concrete slabs of the car-park appears in white; black lines indicate areas of extensive damage.

ing, just three days after the bombing, those commuter trains were operating; Port Authority management and engineering deserve high praise for this achievement.

Our larger responsibility was to decide what portion of the damaged or destroyed structure was to be removed and when, all coupled with the sequence of the restoration/demolition/reconstruction and the temporary bracing of the basement columns against collapse. We had occasional skirmishes with various law-enforcement groups, all of which were solved amicably; our relations with the construction workers were wonderful.

Our major efforts were very large and complex; as well, there were other efforts that were simple, but demanding of time. To list them all would require all too much space and boring text. For a trivial example, all or much of the telephone conduit from lower Manhattan passed through the basement of the World Trade Center. With the blast, many of these conduits fell onto what was left of parked cars. The vehicles needed to be removed in order to proceed with the rebuilding, but first

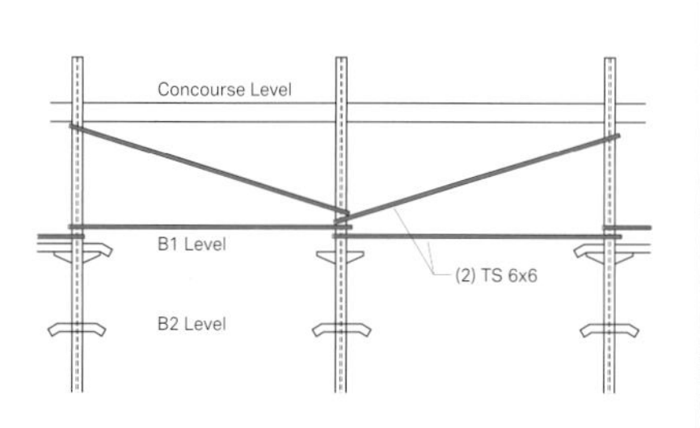

New lateral bracing was needed for the columns that supported the underground garage and the hotel above. On the same day as the explosion, SawTeen and I developed the temporary bracing system shown in the sketch. The photograph shows this temporary bracing in place.

Right Partially collapsed slabs and mangled automobiles among the wreckage.

the conduit had to be raised. Several of our engineers, headed by our Monica Svojsik, worked all night sketching the required details for the raising of the conduit. The field operation took some days. With sound engineering, augmented by Lady Luck, we were told that not a single telephone wire was broken.

As a personal matter, I photographed the license plates of many of the partially-demolished cars, mailing the photos to their owners. While I am almost certain that my photographs are in the dens of many a car owner, perhaps I should say "ex-car" owner, not a single "Thank you!" arrived in my mail.

On one occasion a police car arrived at our New York apartment to take us to the site as swiftly as possible. It was a harrowing experience, driving down the wrong way on one-way streets and the like. Our neighbors, of course, assumed that we had performed some dastardly deed, and had been arrested.

Recognizing that thousands of people were concerned, and that I was the one most qualified to assure the public of the safety of the towers, I accepted multiple TV interviews, always providing an up-beat but honest appraisal of the situation.

While destruction and collapse of the basement areas outside of the towers was extensive, totaling some $400 million, structural damage within the two towers was minimal. Some braces were damaged or destroyed and hollow columns were dented (in many cases, this was 4-inch steel plate); within the North Tower, severe damage was sustained by only one of the diagonal braces and by some internal floor framing I brace as a sculptural reminder for our Connecticut home but there was no practical way for me to transport it nor could I justify spending the time required to navigate administrative hurdles. Much to my delight and amazement, and unbeknownst to me, the foreman had the piece of steel loaded onto a truck, dispatched to our home, and erected in our garden where it stands to this very day, a beautiful testament to the depth and the quality of the work accomplished by the engineers of our offices.

Gil Childers, the federal prosecuting attorney charged with the criminal case against four of the seven alleged conspirators in the bombing, asked for my assistance. I had some graphic panels assembled in our offices, all for purposes of explaining to the judge and to the jury something of the forces generated by the explosion and the nature of the resulting damage. It was all very informal, with no instructions to me, or rules of order, nothing from the federal attorneys, no review before the trial.

On the morning that was to begin my testimony, I dressed as sedately as I was able, grabbed my laser pointer, and appeared in the offices of the attorneys. Following a short wait, I was led to an area outside of the courtroom, and was patted down by a government agent, who discovered my laser pointer. It was an arduous task but I was able to convince the assembled lawyers and police that the laser pointer was harmless.

Once in the witness stand, there were what I presumed to be the usual questions regarding the relevance of my testimony. I got off to a terrible start, almost immediately, by answering the question "What was there in the basement that was capable of creating the destructive force." I responded: "There was nothing in that area that could have created the resulting havoc ... but for a bomb." The defense attorney was all over me because, as a point of law, no evidence had been presented to substantiate the existence of a bomb. Judge Duffy kindly intervened on my behalf with the observation that I was not expert in legal procedures.

Then, the first of our graphic panels was presented. With the aid of my laser pointer, I started to explain the material when Judge Duffy intervened: "What is that pink dot?" I explained that it was just a laser pointer, shortly after which a recess was called. As we reached the hallway, Judge Duffy took me by the arm, asking to inspect the laser pointer. He acted a bit like a little boy with a play pistol, enjoying pointing the light beam at various objects; clearly, to my amazement, he had never before seen such a device. Why, I wondered, wasn't a laser pointer standard equipment in a such a high-profile federal courtroom?

Of the defendants, only one appeared to be involved in the defense. Following closely my testimony, he communicated constantly with his attorney who, in turn, asked me relevant and not-so-relevant questions, with the majority in the not-so-relevant category. Yes, those charged were deemed guilty. Though I knew nothing about the defendants, perhaps my two days in court contributed something to that verdict.

From my experience in the courtroom I concluded that this federal court was in dark ages of the technology, and that the federal attorneys (at least those who dealt with me) lacked fundamental communication skills in the graphic arts. On the other more positive side of the coin, the system worked; the attorneys obtained convictions. Perhaps of more importance, the trial seems to have driven home to other potential terrorists that the stalwartness of the structural systems of the towers was more than sufficient to withstand a mere truck bomb...that a much larger weapon would be required, perhaps leading to the untoward events of September 11, 2001.

There are so many stories to tell, all too numerous for this little book. Following the essential completion of the repairs, with all of our staff, having worked long hours, seven days a week, we rented the good ship Majestic for a wonderful celebratory family dinner and cruise on New York Harbor. Yes, we were justifiably proud of our original ideas and designs for the twin towers and also of our contributions to the achievement of the repairs.

Only two of the major steel members, bolted together and welded to the perimeter columns, were seriously damaged by the bomb. One was blown into the North Tower while the other ended up in the rubble below, cracked by the bomb blast. The latter, taken from the rubble and installed at our weekend home in Connecticut, proudly stands in sculptural testament to the strength and the resiliency of the steel work.

September 11, 2001

Previous

A photograph taken by SawTeen conveys something of the heroism of the rescue workers toiling in a bucket brigade, removing rubble and searching for persons who might still be alive.

When the World Trade Center towers were completed, they stood proud, strong, and tall. Indeed, with little effort the towers had shrugged off the 1993 efforts of terrorist bombers to bring them down. The human-related events of September 11, 2001, however, are not well understood by me . . . perhaps, such events cannot be fully understood by anyone. Without question, the towers fell because the two Boeing 767 aircraft, each laden with more than 10,000 gallons of jet fuel, crashed into them. While a variety of studies have shown that it was the fires, not the aircraft impact that brought down the towers, this book is not the place to examine the modes of failure. Accordingly, herein, I will simply state some matters of fact as I believe to be the case.

The events of September 11 ended the lives of almost 3,000 people, many of them snuffed out by the destruction of the structures designed by me. The damage created by the impact of the aircraft was followed by raging fires, which were enhanced by the jets' full fuel supply for cross-country travel. The temperatures in the towers, in areas at and above the impact zones, must have been unimaginable; none of us will ever forget the sight of those who took destiny into their own hands by leaping into space.

It would seem that about 25,000 people safely exited the buildings, almost all of them from below the impact floors; almost everyone above the impact floors perished, from the impact, from the subsequent fires, or from the eventual collapse. The structures of the towers were heroic in some ways but less so in others. The buildings survived the impact of the Boeing 767 aircraft flying above their rated speeds, an impact very much greater than had been contemplated in our design against a low flying, slow flying Boeing 707 lost in the fog and seeking a landing field. It could be concluded that the robustness of the towers was exemplary. At the same time the fires raging in the inner reaches of the buildings undermined their strength. In time, the unimaginable happened: wounded by the impact of the aircraft and bleeding from the fires, neither of the towers was able to stand long: 1 hour, 43 minutes for the North Tower and 56 minutes for the South Tower (the differential due in part to the location at which the impact occurred).

While in my heart I know that I had no professional responsibility for the design of the stairs and the other exiting systems for the project, the mind is not so kind. Yes, while I had no contractual responsibility for these fire-associated systems, for all subsequent projects I have looked with some care at the architectural drawings depicting these life-safety systems. As an aside, it is noteworthy that each of the two towers had three fire stairs while only two stairs were required by the Building Code.

These charts demonstrate conclusively to me that we should not and cannot design buildings and structures to resist the impact of the aircraft of the future. Instead, we must concentrate our efforts on keeping aircraft away from our tall build-

Above Taken immediately following the aircraft impact, these photographs, taken from our nearby offices, reveal the damage to the east and the south faces of the South Tower.

Below The charts compare the energy of the jet fuel and the energy of impact of various fast-flying aircraft.

Combustion Energy of Fuel

Oklahoma City Bomb ≈ 192 liters

B-25 Mitchell Bomber 2,000 liters

Boeing 707 56,800 liters

Boeing 767 75,700 liters

Boeing 747 216,800 liters

Airbus 380 310,000 liters

Kinetic Energy at Impact

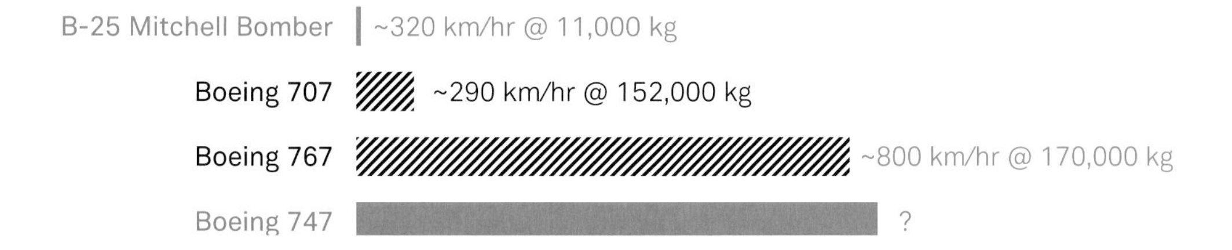

The stark reality of the rubble, along with clear signs that this was a dangerous place for the rescue and recovery workers, demonstrates something of the grandeur and stalwart beauty of that which remained.

ings, sports stadiums, symbolic buildings, atomic plants, and other potential targets.

The extent of the damage created by the impact of the aircraft is almost beyond comprehension. We did not design the superstructures of Building 3 or of Building 7. Both the North and the South Towers were totally destroyed, leaving behind utter chaos surrounded by gnarled naked structural steel. The portions of the two towers that remained standing were painful beyond belief to witness, but in other ways, they were, to me, tragically beautiful.

Building 3, the Marriott Hotel, collapsed down to the structure transfer level designed by us. Fortunately, the people who sought refuge in the hotel lobby survived because of the structural robustness of that transfer level. Buildings 4, 5, and 6 remained standing but were partially collapsed by the falling debris; each burned for about 24 hours. Although there was nothing special about the structural design of these buildings, the remaining steelwork stalwartly resisted the impacts of the wrecking ball. Building 7, designed by others, after burning for nearly 10 hours, collapsed down to the structural transfer level designed by us. The below-grade area under the two towers was almost totally collapsed; in areas surrounding the towers the structures were partially damaged or collapsed.

In my mind, the loss of life and the loss of the buildings are somehow separate. Thoughts of the thousands who lost their lives as the structures crashed down upon them come to me at night, rousing me from sleep; as well, such thoughts come to me at unexpected times throughout the day. Thoughts of those who were trapped above the impact floors, those who endured the intense heat only to be crushed by falling structure, are merged with those who chose to take control of their own destiny by leaping from the towers.

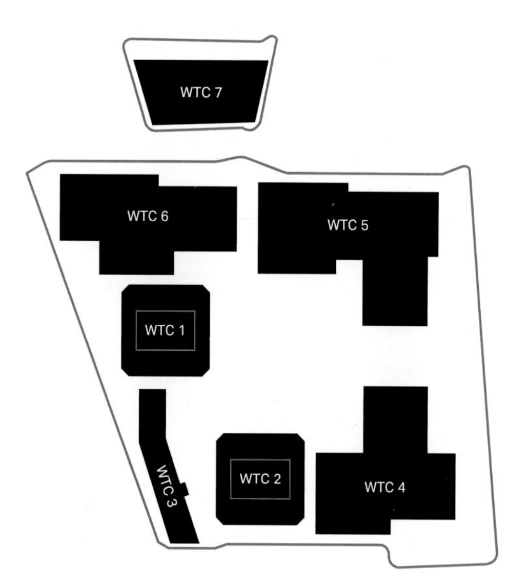

This plan shows the physical location of each of the seven above-grade buildings of the World Trade Center complex prior to 9/11. WTC1 is the North Tower.

The loss of the buildings is more abstract. The towers represent about ten years of concerted effort, both in design and construction, on the part of the talented men and women of our company ... but with the ultimate responsibility falling on my shoulders. It just isn't possible for me to take the posture that the towers are only buildings, that these material things are not worthy of grief.

Yes, we were assaulted with lawsuits, the most being arbitrary litigation wringing money from folks like us to the lawyers. Much to my amazement, the

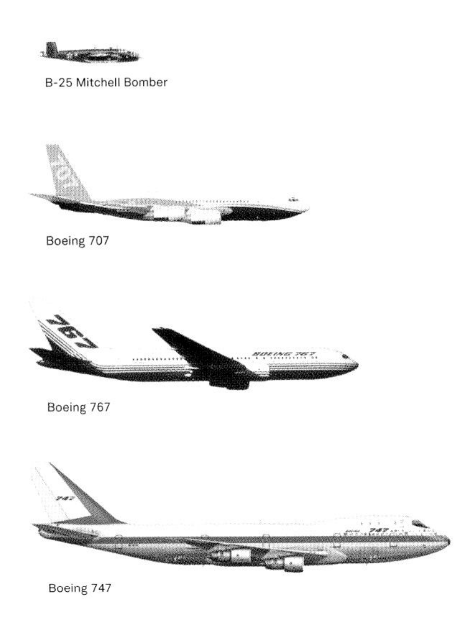

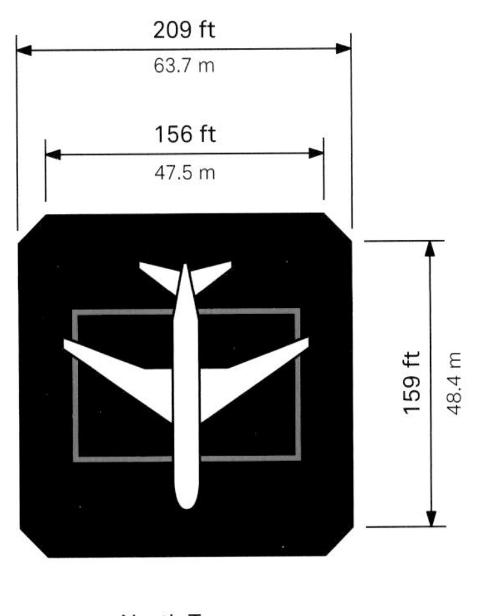

attorneys for Minoru Yamasaki Associates threatened me, personally, for reasons that no sane person could understand. Most, perhaps all, of these lawsuits have gradually faded away...though a new claim could arise at some time in the future.

It would be good to conclude this journey in a positive mode, but it is hard to do so. Some weeks following the event, a very young woman called, asking to meet me, not in our offices but at the public green located at the foot of Manhattan. Her brother had perished in the fires. After exchanging a few words, we hugged each other while crying. An older woman came several times to our offices, even coaxing me into a group therapy session. There were all too many of such episodes, episodes that tug at my heart while demonstrating that I'm not trained to face such events.

I received over a thousand letters, emails, and telephone calls in support of my designs, the collection of which now resides in a set of 3-ring binders to be or already conveyed with the rest of my papers to the Bancroft Library of my alma mater, the University of California at Berkeley. The poignant letters from those who survived the event and from the families of those who did and did not survive, cannot help but bring tears to my eyes. Just a few examples of such letters appear on the following pages.

Do these letters help? In some ways they do; in other ways, they are constant reminders of my own limitations. In essence the overly laudatory comments only heighten my sense that, if I were as farseeing and talented as the letters would have me be, the buildings could surely have been even more stalwart, would have stood even longer . . . perhaps allowing even more occupants to escape. On the other side of the coin, those letters, coupled with the realization that I had accumulated a great design team, a team that bolstered my own weaknesses, provided much of the barrier between me and the abandonment of my career in structures.

Top left The comparative sizes of various Boeing aircraft, including the B-25 Mitchell Bomber that struck the Empire State Building in 1945.

Top center and right On a different scale, how a Boeing 767 relates to the footprints of the towers. The impact to the North Tower was centered while that to the South Tower was east of center.

FROM A PROFESSOR IN PITTSBURGH

Dear Les,

I have tried to write this card so many times in the last two weeks, but I have always failed in my attempts. What can one man say to another in times such as these when we are confronted with such naked aggression, violence, and killing of innocents?

I was teaching class on the morning of 9/11/01 as the life shattering events unfolded. I honestly felt that I would die of sadness as I heard the reports of collapse & tragedy. My mind raced back to the conversations we had almost a year ago in Pittsburgh where we explored similar scenarios to these.

As the days and weeks passed, it became clear that one man's foresight, in the late sixties, led to an elegant structural form; at once able to float on water & survive the slings & arrows of Man's hatred of Man…for a time. Nothing can survive the heat of that fire. Les, you are the unsung hero of this story. Your beautiful buildings saved tens-of-thousands of lives by performing as they did…magnificently. Even in death, their graceful form spared so many as they withdrew upon themselves.

Thank you for thinking the thoughts that saved the lives of so many mothers and fathers when you were expressing art in mathematics a quarter of a century ago.

Maybe some will ask why I thank you now when I was not touched by the tragedy directly. The answer to the question lies in the fact that I know why things were not worse & I am truly, truly honored to know his name & count him among my friends.

With warm personal regards.

FROM THE CHANCELLOR OF A GREAT UNIVERSITY

It is with great pride and pleasure that I write to you, one of UC Berkeley's most distinguished alumni.

I recently came across an article from The New Yorker highlighting your remarkable accomplishments as a structural engineer. In the aftermath of the devastating attacks on one of our country's most cherished landmarks, your work and efforts serve as motivation to celebrate the thousands of lives spared on September 11.

Despite the devastation facing our nation, it is my sincere hope that you are able to reflect upon your role in the World Trade Center's construction, as do I, with great pride.

Our campus is honored to claim you as one of our own.

Fiat Lux

FROM A STRUCTURAL ENGINEER'S PUBLISHED LETTER:

Dear Editor,

Engineers from all over NY, the US and the world have responded to this tragedy from the first day with enormous energy and resourcefulness, in all the rescue and recovery operations. We have the best talent in the world now on site. That is a story that needs to be told someday and will.

But first I believe the nation should know and understand that would it not have been for the intelligent foresight of the WTC design engineer, Leslie Robertson, thousands more would have lost their lives. At his initiative, with the support of the Port Authority, Mr. Robertson provided in the structural design sufficient strength and redundancy to withstand the impact of the largest jet of the time, the Boeing 707. To my knowledge no other building in the world had been designed this way. Because of this, the towers stood 89 and 51 minutes after the impact before collapsing. I heard of one woman who started evacuating the first tower from the 103rd floor after the impact and reached the 56th by the time she heard the second impact. That is 47 floors in 18 minutes. It is clear that many thousands saved their lives in good part because of Mr. Robertson's engineering.

In all of this I was reminded that, as the sun dropped below the horizon in New Jersey, the sunset shadow moved up the sides of the World Trade Center at the rate of one floor per second, creating wonderful sparkles as it rose. It took nearly two minutes for the sun to set on the buildings and almost another forty seconds to clear the top of the TV tower. Now, there are no twin towers to shadow, to sparkle . . . only memories steeped in ruins.

How fragile are the emotions that lie within me! Now, more than a decade later, I can laugh, I can write, I can dream, but I cannot escape the images of the 11th of September. My sense of grief and my belief that I could have done better continue to haunt me. Perhaps, had the two towers been able to survive the events of 9/11, President Bush would not have been able to project our county into war. Perhaps, the lives of countless of our military men and women would not have been lost. Perhaps countless trillions of dollars would not have been wasted on war. Just perhaps, I could have continued my passage into and through old age, comfortably, without a troubled heart.

There is no doubt that one could have made the towers braver, stronger. The fine line between needless conservatism and appropriate increases in structural integrity can only be defined after careful thought and consideration of all the alternatives. But these decisions were made by me in the heat of battle and in the quiet of my dreams. Perhaps, had there been more time for the dreaming. . . .

In conclusion, the events of September 11 have profoundly affected the lives of countless millions of people. To the extent that my structural design of the World Trade Center contributed to the loss of life, the responsibility must surely rest with me. At the same time, the fact that the structures stood long enough for tens of thousands of persons to escape is tribute to the many talented engineers of our company who spent endless hours toiling over the design and construction of the project, making me very proud of our profession. Surely, we've all learned the most important of all lessons: the sanctity of human life rises above all other values.

Examining a small portion of the mangled structural steel temporarily stored in New Jersey after removal from Ground Zero.

ALTERNATIVE THEORIES

There have been so many theories advanced, most attempting to show that it was a conspiracy that led to the collapse. The most prevalent seem to lay the matter on the doorstep of our federal government; this is no place for such discussion.

In lecturing to various audiences, particularly high school and college students, the sensational nature of these theories invariably surfaced. My standard response was, and still is, to be sympathetic to the need for open discussion but to conclude with a simple thought: "Do you really believe that our government has the smarts to pull off this event, undetected, without disclosure?"

In 2010, when I. M. Pei received the Gold Medal of the Royal Institute of British Architects, SawTeen and I decided to travel to the award ceremony in London. It was all very formal, Eileen Pei and SawTeen in lovely evening dresses while I. M. and I wore tuxedos. Following the event, the four of us walked to the curb to hail a taxi. Much to my astonishment, three youngsters approached us to distribute literature for an alternative theory of the collapse of the World Trade Center. They did not know who we were; of course, we did not enlighten them.

MY PERSONAL EXPERIENCE

At the fated hour, I was in Hong Kong, dining with friends. As is the habit in Hong Kong, mobile phones are commonly placed atop the tablecloth. One of them rang. The owner listened, said a few words and turned off the phone before reporting "An airplane ran into the World Trade Center." Having often watched small planes and helicopters well below the tops of the towers, I was not particularly concerned. A short time later, SawTeen called to say that a second plane had struck. Abandoning dinner, I ran to my room, turned on the television . . . and was devastated. Both buildings were on fire; there was no indication of the possible count of persons in the buildings. I could not distinguish between live broadcasting and re-runs. My mind was in chaos.

I focused on getting back to New York, but realized that the press would be after me. The phone in my room rang constantly with reporters seeking an interview. Fortunately, hotel management was able to shut off my phone. I changed the airline that was to take me to Tokyo, got a little rest and headed out through the kitchen so as to avoid the media in the lobby. I don't know whether I did or did not settle my hotel bill but, for sure, I did not do so through the front desk. I took the train to the airport, veering away from reporters while talking with a perfect stranger.

Upon arrival at Tokyo/Narita I took a hotel room almost inside of the airport boundaries. The trick was to find a plane to New York . . . but there were no such flights, as all had been grounded. After several days and several false starts, United Airlines called to say that there was a coach seat available. They said that Governor Pataki of New York had given me top priority, though I found it difficult to believe that such could be the case. Arriving very early at the airport, I went to the United Airlines Club, settling in to wait. Two or perhaps three men approached, identifying themselves as employees of United. They asked for my boarding pass. I assumed that I had been bumped but, instead, they handed me an upgrade to Business Class. Relieved, following profuse expressions of gratitude, I again settled down to wait. Three other men next approached, also asking for my boarding pass but, this time, they upgraded me to a First Class seat, which I gratefully accepted. The plane departed on time (but, of course, there was no schedule), arriving at JFK airport late Saturday evening. The rest has all been reported in the press. Needless to say, passage through the site the next day, and on every subsequent occasion, was a devastating experience, requiring the rigid control of my emotions.

There are those who ask of my feelings toward those who participated in these unimaginable acts. They seem to believe that I must carry a deep hatred for them and for everything that they represent. For a variety of reasons, this is not the case. It is clear to me that the perpetrators of these events carried a deep hatred for that which is representative of the values and the life style to be found in United States. Their own beliefs and their actions are founded in a society that I don't understand. While bewildered by those who perpetrated these acts, who had lived in our midst, seen our homes and our children, I cannot find in myself a hatred for them. I know from my life experience that those who designed and executed these acts do not carry an understanding of the Koran that captured my heart while at Berkeley. Having survived World War II, in some ways I am beyond hatred, beyond fighting. In the words of Chief Joseph of the Nez Perce native American tribe: "I fight no more forever."

In June of 2002 I wrote a long letter to my children. I share a shortened version despite the fact that it contains many imperfections.

This letter, composed on a flight from Dubai to London, is going to be a hodgepodge of disorganized thoughts about Islam, and about those of the Muslim faith…thoughts stimulated mostly by the unimaginable events of 11 September and thoughts stolen from a variety of sources. I'm not well-versed in these matters… so that you will likely detect many instances of my imperfect understandings. This is my personal time for reflection...with the cherished time for this letter being a part of the vehicle for that reflection.

The foundation for the thoughts expressed herein is that today's America is not as beloved as was yesterday's. Many are not concerned...having not determined why or even wondered why we've changed. Somehow, during the excruciatingly painful and mind-opening period of World War II, God passed from my vision. However, today, being an "I'm Completely Bewildered" kind of guy, I'm not a Buddhist, a Christian, a Hindu, a Jew, a Muslim, or an atheist; perhaps I'm an agnostic, perhaps not. It was an Imam who once said, "There is only one thing less likely than the existence of God, and that is that He does not exist." Within me, somewhere hard to find, there is a level of agreement with that concept.

Many believe that I do or that I should carry a deep hatred for Muslims in general and for those who twice attacked the World Trade Center in particular. But, if there is that hatred, I just can't find it. On balance there are those remembrances; the bombs on Hiroshima and Nagasaki and so many others…all heinous acts of 'civilized' people. Were the perpetrators terrorists?

And, were the Japanese Kamikaze aviators being "terrorists" when they crashed their planes into our ships? Or were they patriotic zealots, freedom fighters, attempting to protect their homeland? While it was just such a pilot who killed my friend, I'm not able to "hate" the pilot or knowledgeable or saintly enough to judge such questions.

The totality of these thoughts creates in me a sense of despair, but not of hatred.

Somehow, our President Bush has taught us to paint Muslim activists as "terrorists." While the vast majority of the Muslim world thinks with horror on those who planned and accomplished the events of the 11th of September, many of them believe that those activists are misguided "freedom fighters"...very much the same as the population of Japan thought of their Kamikaze aviators. How are we to learn the difference, or is there a difference?

Clearly, Bush would have us believe that there are two kinds of wars and two kinds of rhetoric…those by the other fellow that are bad, and those by George W. Bush that are good. Are we to learn of the differences between "terrorists" and "freedom fighters" from him?

Both the President and the US media have wrapped themselves in the American flag and love of country; but now this symbol of American democracy is affixed to military means of destruction. George W. Bush has promised the American people a long war and, for purposes of re-election (my opinion), seems to be bent on making it so. The operation in Afghanistan was called "Enduring Freedom"...but it's not "enduring" for those who have and who will continue to perish in these conflicts.

By now, we all know that madrassa is the Arabic word for "school." Contrary to the reporting of some of our media, with few exceptions, Arabic-language schools, madrassas, do not teach their students to hate the West or to blindly follow Islam; these are not, as some would have us believe, finishing schools for terrorists.

No person who loves peace or who respects the rights of others could have participated in the terrible crimes of the 11th of September. We should not and must not hold accountable for these events those who practice the peaceful faith of Islam.

As do all of the great religions, Islam is a religion that teaches tolerance, justice for all, and respect for human rights; it's a religion that promotes peaceful co-existence, and recognizes the dignity of man...but Islam does so in ways imperfectly understood by me.

True believers in freedom despise only those who practice or condone murder, not those who adhere to the beautiful religion of Islam. It's intolerant of us to tar those who follow Islam with the brush of fanaticism and hatred. Don't all religions have their fanatics?

There are those who seem to believe that Islam promotes hatred and fanaticism while, as is the case for all of the great religions, Islam's foundation is to be found in peace. The very name "Islam" is rooted in the Arabic root word "salaama" which means "peace." Acts that are contrary to peace, such as the murder of innocent people and of suicide, are not sanctioned by Islam. Indeed, Islam advocates and supports human rights, freedom of action, and the right to choose...but all within an organized social framework, based in part on justice, truth and charity...a framework not understood by me.

In the Americas, in Asia and in Europe, there exists a kaleidoscope of cultures within each area…each speaking different languages and each having different customs. And so it is with the Middle East. My, how the complexities of history and language are so beautifully intertwined!

Those of us from the West allow ourselves a unique cultural identity, with each nation seen as a separate entity, practicing its own mix of cultures and beliefs. Why do we seem unable to see the Middle East as culturally diverse as are other regions of the world?

Seldom do we recognize, much less understand, that countries throughout the Middle East are striving toward the building of social and economic stability. Bahrain, Kuwait, Lebanon, Qatar, Syria, Tunisia and the United Arab Emirates, particularly, are to be lauded. Perhaps, if we all tried harder to understand and appreciate the variability of the Arab culture and society, a more sensible foreign policy could be derived.

At the same time there are the terrible conflicts between the Palestinians and the Jews, there is Saddam Hussein, there are the US sanctions against Iraq and our military presence in Saudi Arabia, and there are some corrupt Arab governments (as there are all too many in Asia and in the West). We need to sort out a way of resolving these issues; that way can be found only through the peaceful intervention of the United States and of Europe. Indeed, there is no benefit in war and no rationale for our present policies in the Middle East.

The Torah, because of the lack of revision from the time of the early Christian epoch, renders Muslim scripture as one of the few reliable historical sources for Christian scholars. The unedited nature of the Koran gives Muslims confidence in their understanding of the nature of Jesus, which was as a servant of God, as a Prophet. Neither the Torah nor the Koran declared the divinity of Jesus; it is, perhaps this single difference that has helped created two millenniums of animosity between the Christian, the Jewish and the Muslim worlds.

In short, I believe that the Muslim Koran is a continuation both of the Christian Bible and of the Jewish Torah, with all three carrying the same messages of peace, justice, truth and compassion.

It seems to me that it is believable that idealistic young men and women from another culture might come to or be led to believe that the United States is an evil thing...a thing that deserves to be destroyed. How to dispense our own feelings of love, peace, equality, and the like? erhaps we should roll up our sleeves and get to work making life better for the peoples of the world? Fewer billionaires, and fewer citizens of our world perishing from starvation? We seem to be pretty efficient at waging war in the Middle East, but just can't seem to get it right for the war on AIDS in Africa.

I hope and believe that, if we do all in our power to make peace with the Muslim world, following two or three generations of hard work, we'll be able to live comfortably together in peace, and we'd better get right to it!

And now there is the moment to apologize for the undisciplined nature of the thoughts in this letter. In truth, I had to write to someone; perhaps it's just the thought that you would hear me through with the patience exhibited only by one's own children.

Lotsa luv!
Dad

Meishusama Hall

SHIGARAKI PREFECTURE, JAPAN. 1970

We first encountered Shinji Shumeikai (Shumei), a Japanese religious group, when Minoru Yamasaki was designing Shumei's Sanctuary Building, known as Meishusama Hall (1982), in the remote hills of the Shigaraki Prefecture outside of Kyoto. For reasons never fully understood by me, there had been a falling out between Yama and our John Skilling, Yama's rigid nature resulting in the inevitable cessation of all relations between our two firms. Following an appropriate period, I penned a note to Yama pointing out that the differences had to do with John, but that I remained deeply interested in working with him. In due course, I received a letter: "Les, you are right. At your convenience, please visit with us."

In the meantime, Yama had turned to another US engineer for the design of the Sanctuary Building, but that design had been rejected in Japan, and an internationally famous and talented Japanese engineer, Yoshikatsu Tsuboi, was chosen for the task. It seems that Tsuboi-san and Yama were both strong-willed and stubborn, finding themselves at odds over many issues. Of course, often, they were able to sort out such things, but one area in the roof skylight defied settlement. This came up in the course of my conversations with Yama, but not with him asking for my assistance, and I made no such offer.

As time passed, Tsuboi came to New York, explaining to me the controversy over the skylight detail. He was quite blunt: "You must explain to Yamasaki that I am right, and that he is wrong." While understanding his concerns, I promised to study the matter, but did not promise to take a stand in any direction.

Following Tsuboi's departure, and following my study of both designs, I developed a third solution to the skylight framing, a solution that met and improved upon Yama's aesthetic considerations as well as overcoming Tsuboi's structural objections.

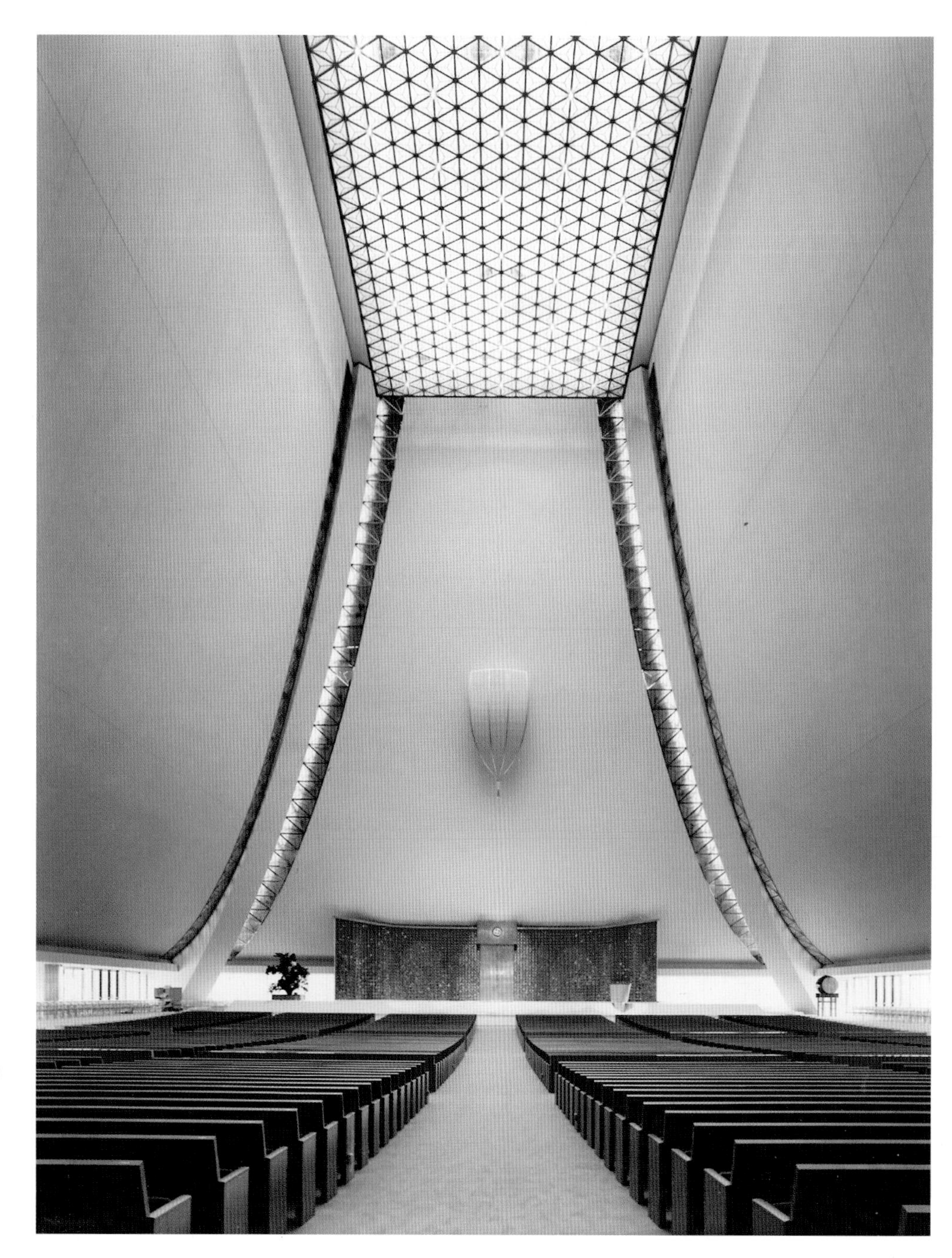

The grandeur of Meishusama Hall is evident in this soaring interior view. The main Sanctuary, capable of holding 5,600 worshippers, is just one part of the overall Shumei complex.

While I was not present at a subsequent meeting of the two, it seems that harmony returned, my suggestion was adopted, and the designs moved forward.

Following the onset of construction, as SawTeen and I were in Japan, Tsuboi suggested that we meet for breakfast before traveling to the construction site. It was a glorious day, with the skies clear and blue, the wind moderate, and the temperatures comfortable. The tour of the site was good fun, while assisting SawTeen and me in the learning of just a little of Japanese construction technologies. Afterward, a huge luncheon was held with the construction staff, with Kaishu-sama (Mihoko Koyama), spiritual leader of Shumei, her older brother "Richard" (English name) Koyama, and her daughter Hiroko Koyama, all seated at the dais together with Dr. Tsuboi, SawTeen, and me. Richard sat in the very center; as the meal was completed, all watched him for a signal. He rose ever so slightly in his seat, causing all to begin to rise. Then, as he sat back down, all others sat down also. This appeared to me a ceremonial display of strength: Richard ran the show.

That night, Richard took over the entire space of a fine restaurant in the hills of Kyoto. The guests were Yama, Tsuboi, SawTeen and me, perhaps one or two others. It was a fine dinner, featuring the renowned Kobe beef...despite the fact that we rattled around in an otherwise empty restaurant.

By this point I had established a relatively close acquaintance with the Koyama family, but with no contractual agreement nor any financial compensation for my efforts. Our friendship continued to grow and, indeed, beginning with I. M. Pei's bell tower in 1989, I became significantly involved in the development of Shinji Shumeikai's world center.

Harrison, Abramovitz & Abbe Architects

While perhaps better known as Harrison & Abramovitz, Harrison Abramovitz & Abbe Architects (HAAA) was the architect for the US Steel Tower, Pittsburgh, which, when constructed, was the largest privately-owned building in the world. The firm had designed some of the most famous buildings of our time; with close ties to the Rockefeller family, their offices were in Rockefeller Center.

Somehow, I met Wallace Harrison only once, Max Abramovitz (in the photograph opposite) more often, and worked closely with Charles H. (Charlie) Abbe. While Charlie was more than twenty years my senior, we fit well together and became good friends. Indeed, even after his retirement, I visited with him. Preferring to be without an office, Charlie worked in the drafting room, sitting on a drafting stool before a drafting table, pencil, tracing paper and T-square in hand.

While Charlie was the designer for the US Steel Tower, he graciously accepted the very significant changes that we proposed (detailed below).

Because of its unique structural systems, the scale of the project, the eminence of the architects, and the fact that it is an all-steel building, following construction, the structural systems of the US Steel Tower received widespread recognition from structural/contractor organizations and press. The architectural award-givers were less enthusiastic.

For most of the black-tie award ceremonies Charlie preferred to be a no-show. SawTeen and I would invite Max Abramovitz and his wife Anita to join us. Since I had arranged the seating, it was easy to converse with Anita, only to discover that they had never been to Japan! Anita explained that Max did not like to travel unless the cost was borne by others.

Almost instantly, my mind focused on a lecture tour to Japan to be hosted by a Japanese company. We shifted our attention to Max, explaining that Japanese companies were desirous of having a lecture series provided by Abramovitz & Robertson. Max was delighted. Now, all I had to do was to find these generous Japanese companies! Later, turning to Isao Kimura of Nippon Steel, I found that the arrangements were welcomed. In short order, Max, Anita, and I were off to Japan!

As it turned out, the assembled audiences in Japan were far more interested in the structural aspect of the design than in its architecture. Max had a significant up-hill battle to enthuse his audiences. Fortunately, I had organized the lectures so that Max spoke first, with my follow-up more responsive to the interests of our audiences, who then departed with smiles on their faces.

Our final lectures were in the fascinating city of Kyoto. We were housed in an incredibly expensive *ryokan*, or inn, each party in a private building, with a personal cook and housekeeper/butler. On our last night together, we dined in the Abramovitz quarters, there to enjoy really fine examples of Japanese cuisine. Over dinner, I regaled Max and Anita with stories of the *ryokan's* community bathing, a sort of river, with the water incredibly hot at the top and just warm at the bottom.

Following dinner, the Abramovitz's maid explained to Anita that her "bath" was ready. As one would expect, refusing the offer, she was totally unwilling to participate in communal bathing. I was able to convince Anita that the entire "river" was for her use alone, and that she would enjoy the experience. As she departed with the maid, it was the last time that I saw her in Japan.

The following morning, I departed early for an appointment, leaving Max and Anita still asleep. When they had finished breakfast, Max attempted to pay for the room and services, only to be told that they did not accept credit cards—an error in communication on my part. Fortunately, I'd left behind the address of my meeting at Takenaka Komuten (now Takenaka Corp.). Leaving Anita as "hostage," Max appeared in our conference room, received the appropriate amount of cash, and returned to rescue Anita.

A week or so later, after returning to New York, I received a call from Harrison, Abramovitz & Abbe's business manager. When we met later that morning, he informed me that our invoice for design services on the curtain wall of the US Steel Tower would not be honored. While the services were completely outside of our scope of work and completely inside of the architect's, we were not paid. Of course, for fear of the adverse publicity and because we considered the Abramovitz's to be our friends, we could not sue for these extra services. Such are all too often the realities of engineers contracting with architects!

In the years following, we did little work with the firm. I have been told that its offices were closed a few years after the opening of the US Steel Tower.

At work on the US Steel Tower.

US Steel Tower

PITTSBURGH, PENNSYLVANIA. 1965

With a gross area of almost one acre on each floor totaling 2.9 million square feet of column-free office space, the 64-story, 841-feet-high United States Steel Tower in Pittsburgh was, upon completion in 1970, the largest privately owned building ever constructed and the tallest in the world outside of New York and Chicago. Having been sold, the building is now known as 600 Grant Street.

The facade is Corten, a weathering steel developed by the United States Steel Corporation and intended to be showcased in this building. In an industrial atmosphere, and with time, this steel develops a tight rust, which generates a rich and handsome patina. Folks in Pittsburgh call the building "The Rusty Triangle."

THE TEAM

There were many players on the design team, the most important being: Galbraith-Ruffin Corporation who was responsible for construction supervision and building management. Peter B. Ruffin led their team.

Turner Construction Company was the Construction Manager and Contractor. The very talented and perceptive Leslie V. Shute headed the team; he contributed thoughtfully and knowledgeably to every one of the weekly project meetings.

The American Bridge Division of United States Steel both fabricated and erected the steel work. Ron Fluker headed their in-house research effort.

Harrison, Abramovitz & Abbe was the architect, having also designed Rockefeller Center, the United Nations Building, and a host of other famous buildings. Charles H. Abbe was the Project Designer of the US Steel Tower.

Our firm, Skilling, Helle, Christiansen, Robertson, was the Structural Engineer for the tower, with me as the Partner-In-Charge and the Project Designer, and Harold D. "Hal" Roet as Project Engineer.

Edwards & Hjorth was the Structural Engineer for the areas outside of the tower, with Alfred E. Danner heading this group.

Though we were never made privy to the name, prior to our becoming involved, another structural engineer had been chosen for the project; apparently, it was the intervention of the upper-echelon executives of US Steel, perhaps driven by our designs for the World Trade Center, for the IBM Building Pittsburgh, and for other projects that resulted in our being chosen for the commission. I do know that the Chairman of the Board of United States Steel had cited five buildings for innovative structural design; three of the five were designed by our firm.

While, contrary to my wishes and without my knowledge, there was an early attempt at intervention on the part of our Seattle office, with the support of US Steel, I had no difficulty in restricting the work to New York. In my view, having developed the designs that had generated our selection for the commission, and having conceived the ideas that made possible the architectural and the structural systems for the US Steel Tower, I was rightfully unwilling to share that recognition with other than those persons in our New York office who made it all possible.

Other major consultants on the team included:

Jaros Baum & Bolles, the building's services engineer (mechanical, plumbing and vertical transportation). Donald E. Ross was the Partner-In-Charge.

Ebner-Schmidt Associates was the electrical engineer.

Meuser, Rutledge, Wentworth & Johnston was the Foundation Engineer, William (Bill) Meuser being the Partner-In-Charge. With ground breaking in 1967, I was not yet 40 while Bill was well over 60. Perhaps because of my age, it was war-at-first-sight, with Bill convinced that I did not have the experience or the talent to undertake the design. The good news is that, with the topping-out in 1969 and initial occupancy a year later, Bill and I buried the hatchet over lunch and a bottle of fine wine. Indeed, some years later, Bill's firm called on me to explain certain aspects of our foundation designs for other projects.

GENERAL REQUIREMENTS

For most buildings, the paramount requirement imposed by the client, and the most important achievement of the building's designers, is the construction of a quality product at a budget price. For the US Steel Tower, additional client requirements included:

- The building was to be a symbol of the technological achievements of US Steel, particularly exemplifying the use of Corten steel in building construction and in putting said steel on display;
- A systems approach to design and function was required for optimum building performance;

- Provision for large, column-free areas with maximum window perimeter (as explained below, this requirement was not met);
- Construction in office areas was to be completely modular, permitting maximum flexibility of space utilization;
- Framing systems were to be organized to accommodate a railroad track passing under the tower, directly under the largest building column known to us;
- Provisions for a rooftop heliport, able to accommodate vertical takeoff jet aircraft when such planes became operational.

Among our most notable achievements were:

• A unique structural system, in essence a series of 3-story buildings stacked within a primary structure;

• Exterior structural columns of weathering Corten steel, devoid of cladding and free of conventional fireproofing, made possible through the introduction of cooling liquid within the hollow columns. I did not develop the concept of the liquid fireproofing system but did authenticate the technology and developed the structural systems that made implementation of the concept possible.

Taken from my own designs for the World Trade Center was a rooftop space frame within a triangulated braced core, which mobilizes both the tensile and the compressive capabilities of the exterior columns. Today, we call these outrigger trusses. Further, the rooftop space frame significantly reduces the building movement in high winds, while acting to control temperature-induced differential vertical expansion and contraction between the thermally controlled columns of the interior service core and the exterior columns which are exposed to the atmosphere. The outrigger trusses eliminated the need for a conventional rigid frame while reducing the floor-to-floor height to only 11-feet-10-inches.

Quenched, tempered, and other high-strength and carbon steels were used to develop truss work, spanning the existing railroad tracks while supporting the tower above. The system was designed so that, while spanning the tracks, the top chord of the trusswork deflects uniformly downward without increased deflection at the center of the span.

Standing firmly on the shoulders of the pioneering research and designs for the World Trade Center, accomplished by Prof. Alan G. Davenport and ourselves, a comprehensive environmental and wind engineering investigation was designed and implemented.

The six models tested in the Boundary Layer Wind Tunnel of the University of Western Ontario had identical stiffness and density, with only the plan shape varying. Comparative results are shown in the graph below, with the circle representing the basic horizontal line at 0.002. The advantage of reduced wind-induced lateral response coupled with the attaining of six corner offices was enough to have the "notched" triangular design accepted by all parties.

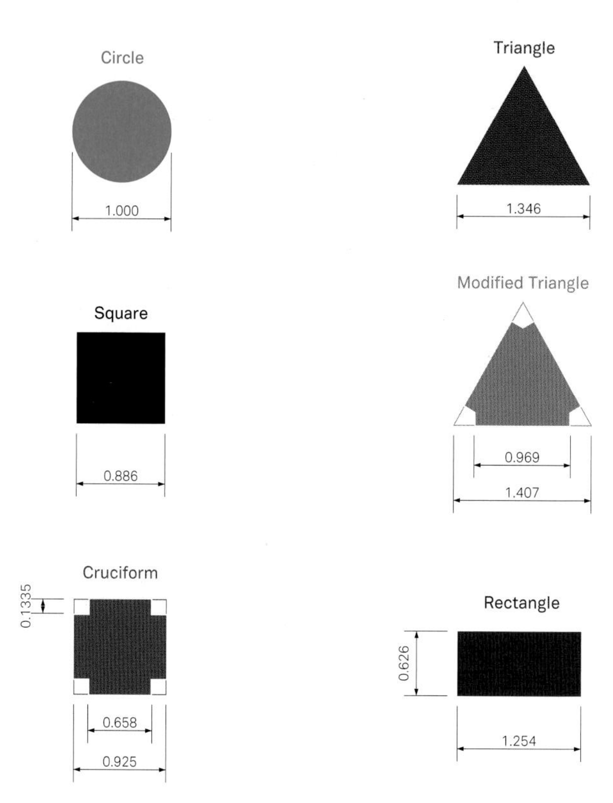

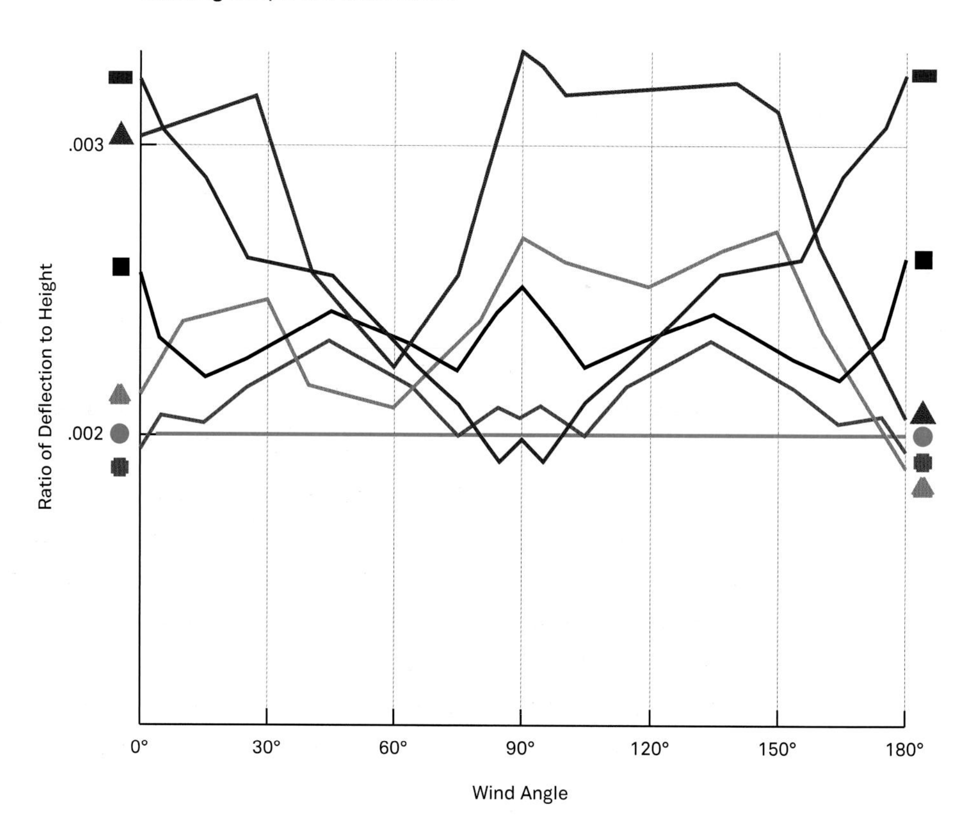

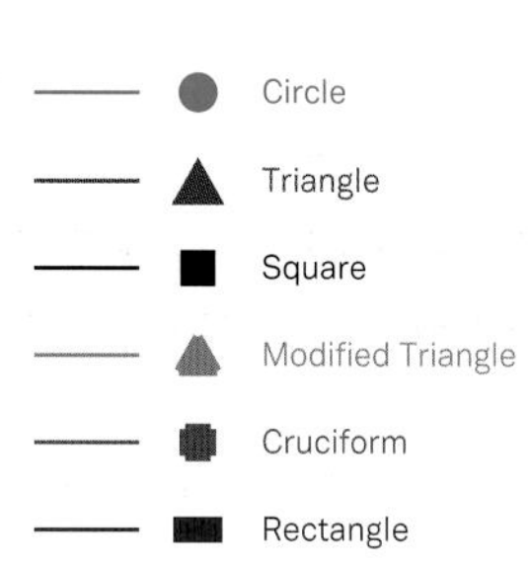

WIND ENGINEERING

While the general topic of wind engineering is discussed with the World Trade Center in this book, it is important to remember that the technology was just evolving and that the twin towers and the US Steel Tower were the first ever to use wind tunnels to gauge wind loads and human comfort in tall buildings, a practice which quickly became an industry standard. In the case of the US Steel Tower, the question of aerodynamics was important in determining the shape of the building.

The six aeroelastic wind tunnel models, with our preferred modified triangle enriched with a bit of color.

Initially, the architects had intended a building defined as a pure triangle in plan. Having come to understand and appreciate the unsteadiness of a three-legged bar stool, I was skeptical. Further, I had concerns regarding the swaying motion response of a purely triangular building in a turbulent wind.

To examine the structural characteristics of the building, I proposed that six plan shapes be tested in the wind tunnel, each with identical mass and stiffness properties, and a common gross area: a circle, a triangle, a modified triangle with notched corners, a square, a modified square with notched corners, and a rectangle.

The rectangle was the shape that Galbraith-Ruffin, the developers, believed to be the most effective—from the perspective of the rental agents. The architects preferred the pure triangular form.

We took the architect's preferred triangle but modified it with notches in the corners to enhance the aerodynamic damping, thus reducing wind loads and wind-induced sway. I took the liberty of enhancing the appearance of this model. The other shapes (square, rectangle, and circle) were tested in order to provide a basis of comparison with comparable buildings of the time.

In the diagram opposite, we see the six wind tunnel models. Each was a stiff (non-bending), simple model able to deflect laterally, pivoting in flexure and in torsion about the base. The diagram shows the response of each of the six shapes, the sum of the steady-state and the dynamic (oscillating) response. As expected, the circular shape produced the lowest side-to-side lateral movement under the action of turbulent wind. Next is the cruciform (modified square), followed by the pure square. Our modified triangle is next. Then comes the pure triangle with a big increase in wind-induced response associated with the manner in which the air flows around the triangular form. The rectangle, because of its high ratio of height to width in the short dimension, was the worst-performing of the lot.

The modified triangle, with its much improved performance over the pure triangle, was selected. The decision was based, I am sure, more on the advice of

Pete Ruffin, the developer, and architect Charlie Abbe, than on me. Of course, Max Abramovitz may have been involved with the decision but, for whatever reason, I seldom worked with Max. While I could be wrong, I believe that they chose our shape over more efficient shapes because it was visually more striking than the others. At that time, perhaps understandably, I celebrated the choice and moved on to other issues.

Clearly, for the US Steel Tower, with important structure open to heating/cooling from the atmosphere, it was not appropriate to combine the very highest wind speeds with either the lowest or the highest recorded air temperatures. One of our engineers, Dr. Peter Chen, and I accomplished a joint probability study, evaluating the appropriate combinations of wind speed and air temperature. This analysis, perhaps the first of its kind, provided the basis for the design-level wind and temperature loadings for the US Steel Tower.

Commonly called a "pressure model," it is used to obtain both the steady-state and the fluctuating wind-induced pressures and suctions on the metal and glass on a building's facade. With the cost of the facade often exceeding that of the entire structural system, it is essential that safe and reasonably conservative criteria be developed for its structural strength. The same model was later used to evaluate the environmental wind speeds, which are important in examining the comfort and the safety of pedestrians in the immediate area of the building.

The wind tunnel model with one face removed to display the interior equipment. The small plastic tubes travel from a pressure tap (basically a small opening in the facade) to a pressure transducer.

In the technology of that time, about a thousand pressure taps could be implemented and measured at one time. Where more than a thousand pressure taps were required, the process would simply be repeated for the remaining pressure taps.

Back in the late 1960s, US Steel had the largest non-commercial airline in the world. The roof of the headquarters, then, was not designed for the puny run-of-the-mill corporate helicopter. Instead, we designed for loads from a much heavier aircraft, a vertical-takeoff jet which was not yet in the marketplace. Searching for criteria, we designed the roof for the collapse load of the landing gear of the Boeing 707 aircraft, the largest commercial aircraft of the time. To assist in imagining the enormous size of this rooftop landing pad, I can attest that I've been a passenger in the takeoff of a DC-3 from a smaller field—JATO (jet-assisted), of course.

Then there was the issue of turning vanes, which were intended to redirect the vertical flow of storm-driven air, forcing it to flow smoothly over the rooftop of the build-

An elevation of one of the three sides of the services core with the differing grades of steel denoted by color. Since high-strength steels are stronger, for the same load they are smaller in area, and more flexible and hence will shorten more under loading. The depicted system is "softer" in the center, forcing gravity loads to move toward the edges of the trusswork. This is essential because of the presence of the rail tunnel passing through the building.

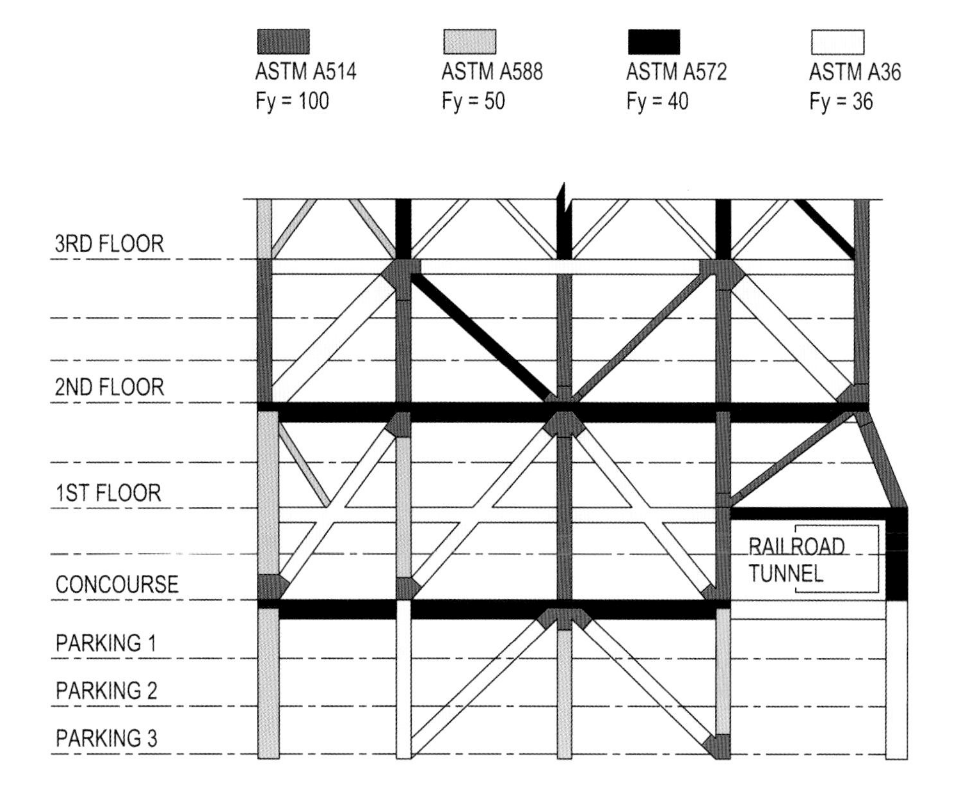

Cross-section through a primary floor spandrel showing the cladding, the fire-resistive material, and the column supporting the two floors above.

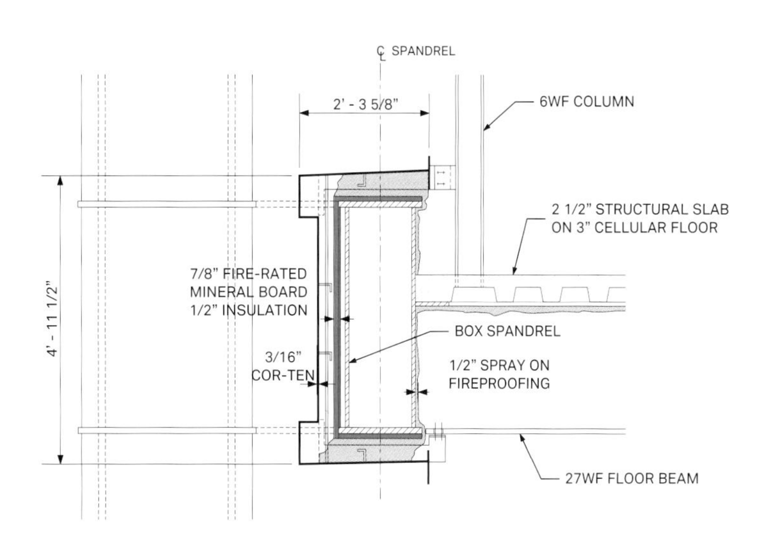

Fundamental Dimensions of Spandrel Cover

ing. An important precedent was New York's Pan Am Building, which had turning vanes at the edge of the roof and the only heliport then in operation atop a high-rise building. Ever skeptical, I suspected that the turning vanes were far too small to handle the large quantities of air flowing up the vertical face of either the Pan Am or the US Steel Tower. Accordingly, we tested a model of the upper areas of the Pan Am Building to show, conclusively, that the turning vanes were not effective.

To complete the story, a model of the top of the US Steel Tower was tested to again show that the turning vanes were not effective. The turning vanes on the Pan Am Building had cost almost $750,000 in 1970, well over $2 million today. My intuition saved our owner/developer an amount well in excess of the entire fee paid to us for the structural engineering but, of course, we are expected to, and do, save clients monies commonly in amounts well in excess of our fees.

Recognizing that, with the perimeter columns completely outside of the building facade, there must be structural steel passing through the curtain wall. That passage had to accommodate both the structural and the architectural construction tolerances.

The architects proposed that these exterior columns be spaced at 13 feet on center, a sensible spacing for the usual perimeter columns. I argued that, because of the cost of the column/facade interface, spacing the columns at 39 feet on center would both speed the construction and reduce the cost. Since an increase in column spacing results generally in an increase in cost, not a decrease, while there was considerable debate, all parties concurred with my proposal.

Taking yet another step, we proposed that, instead of connecting the perimeter columns to each floor, the columns should be connected to the building at every third floor, a "primary floor," with the two intermediate floors posted down to the primary floor on 6-inch interior steel columns. From the architect's original concept of columns at 13 feet on center to this megastructure concept, the number of penetrations through the curtain wall was reduced by a factor of almost nine, and the melding of the architectural and structural tolerances enormously simplified.

I have been able to partially compensate for my fear of heights, enjoying lunch at the US Steel Tower while sitting on the very edge of an upper secondary floor, with my legs dangling over the abyss below. Sometimes I was able to convince a guest to sit beside me. I would then point out that the floor above, the primary floor, was firmly affixed to the exterior column but that the floor on which we were seated had no such connection. I would then explain in jest that the floor cantilevered from the services core. Not one of my guests had the ability to peer over the edge to find the 6-inch wide-flange column below, and some may believe to this day that the floors do cantilever the 52-foot span.

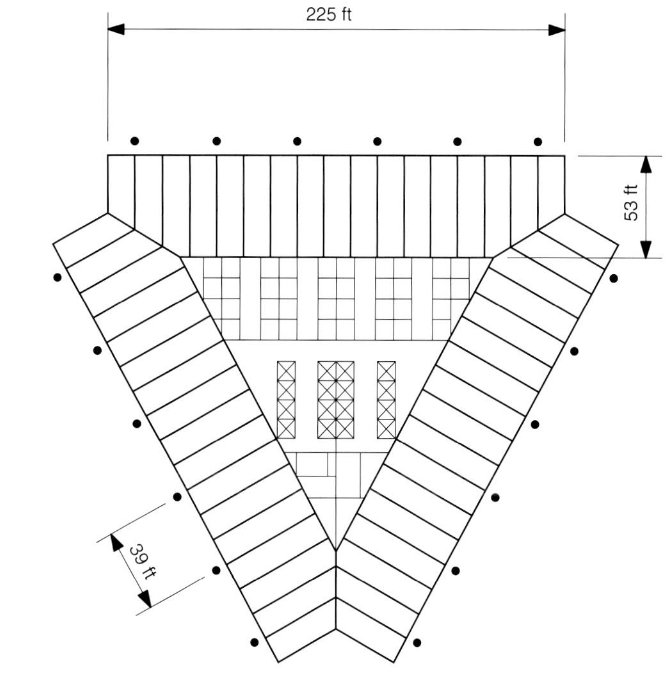

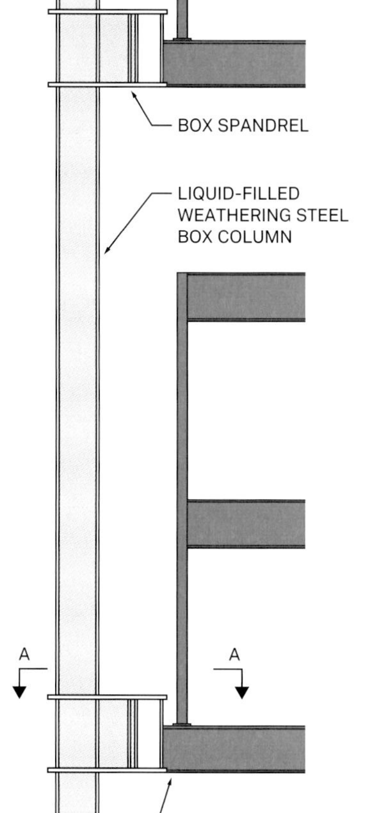

Fire Protection – Liquid-Filled Columns

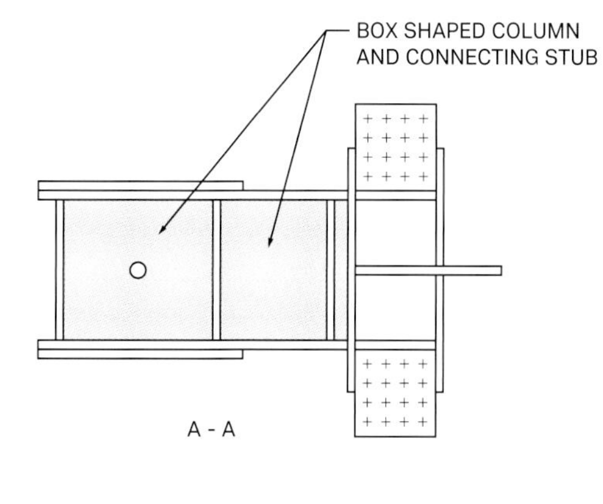

Top The plan view shows the typical framing as well as the location of the perimeter columns.

Far left A depiction of the 3-story module, the columns of which are 6-inch wide-flange sections.

Left A plan depicting the connection between the box column (liquid-filling in grey) and the connection to the box spandrels.

The United States Steel Corporation insisted that the building's curtain wall make use of both their Corten steel and a design for which they owned the patents. It was and is my opinion that, aesthetically, this was far from the best choice. An all-glass facade, emphasizing the every-third-floor spandrels and the exposed steel columns, would be in harmony with the structure and with the architecture, instead of hiding these essential features of the building. Unfortunately, my arguments fell on deaf ears and the company's wish was followed.

But wait! The plan shape of the building and the spacing and character of the perimeter columns came from us, the structural engineer, and the curtain wall design came from the client. In essence, the entire exterior aesthetic of the building was conceived by parties other than the architects.

Both for the World Trade Center and for the US Steel Tower, as far as we know, outrigger space frames were incorporated into the structural concept for the first time.

For the US Steel Tower, unlike the World Trade Center, the outrigger trusses were a part of our original concept, serving two essential functions:

First, they provided a significant stiffening of the lateral force system, desirable in the triangular building, to partially compensate for the inherent instability exhibited in a three-legged stool.

Second, they ameliorated the temperature and the gravity-induced shortening and lengthening of the perimeter columns.

Of course, the outrigger trusses could not completely eliminate the differential shortening and lengthening of the perimeter columns with respect to the columns of the services core. This differential results in a small change in the slope of the floors. How small is small? The Chase Manhattan Bank, located just off of Wall Street, while not as tall as the US Steel Tower, has outside perimeter columns. I interrogated the building manager to determine that the change in slopes existed, but not to the extent that it was an issue with occupants or maintenance personnel. Intrigued by the matter, the manager permitted us to run levels on his floors. Coupled with the probabilities of both high and low

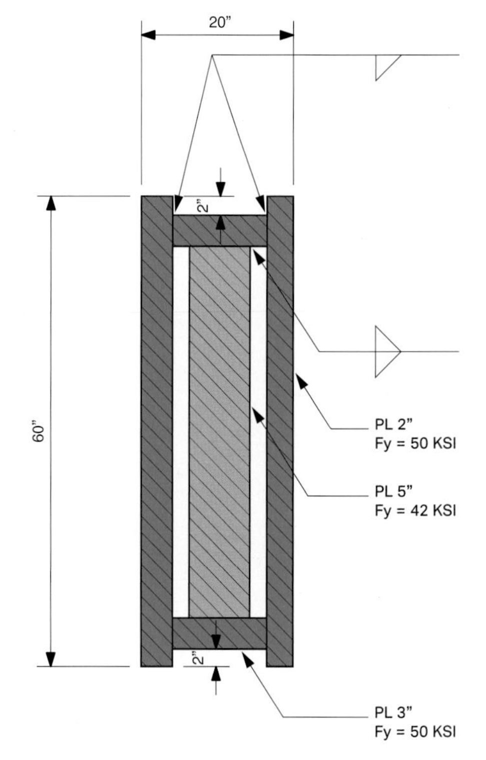

A cross-section of one of the paired columns in each of the three corners of the services core.

temperatures in both New York and Pittsburgh, we were able to establish criteria for the US Steel Tower, a most fortuitous circumstance.

Although not recognized in the building codes, for the largest of the columns, we used a hybrid of mixed grades of steel. The steel in the perimeter of the column has a yield point of 50ksi. The steel forming the inner core of the column is of a lower yield point and lower cost, at 42ksi, the highest strength steel sensibly available in the required thickness of eight inches. It is likely, however, that one can be mesmerized, at least intuitively, into believing that the entire column is of the higher yield material.

The fabrication superintendent proudly displays the fine workmanship achieved in these huge columns.

Of course, we accomplished detailed calculations supporting our design concept. For sure, this hybrid column significantly reduced both the cost and the size of these columns. One of these columns, as seen in the storage yard, offers a sense of the scale of these huge pieces. As well, one can sense the pride showing in the face of the plant superintendent (previous page).

The US Steel Tower was complex and innovative in many respects, notably in fire resistance, but the design concept is deceptively simple:

- The columns and the connecting stubs are fabricated from Corten weathering steel, designed to provide a protective oxidized coating that does not require painting;
- The liquid within the columns, fortified with anticorrosion and antifreeze chemicals, has a weight of about 30 percent more than that of pure water;
- Through piping, at their bases and at the tops of zones, all of the columns are connected together;
- As one or more columns are heated in a fire, the lighter, warmer liquid rises in those columns, bringing cooler liquid from below, traveling upward in the columns being heated by the fire and downward in the remaining columns;
- To maintain the columns full of liquid, surge tanks are provided within each vertical zone to deal with the everyday and the fire-induced heating of the liquid;
- The box spandrels, like the rest of the structural steel, inside of the Corten spandrel, are conventionally fireproofed with a spray-applied material.

Perhaps not obvious, the liquid pressure is rather high in these columns. It may be that this was the largest, and almost surely the longest, collection of rectangular pressure vessels ever constructed. Did they ever develop leaks? In 1971, during construction, the first leak emerged. Passersby on the sidewalk were astonished to find it raining under a clear blue sky. Fortunately, over the years, few leaks have occurred.

While visiting the fabrication works, I was quick to note that the shop welding of the columns and stubs seemed to be strong enough, but that visually the welds left much to be desired. Calling for a meeting of all of the welders and the welding foremen, I explained that these welds would be three or four feet from the critical eye of the executives of their own company, and that their workmanship would be seen by thousands of people. The meeting produced a remarkable change in the visual quality of the welds. Not only did the work improve going forward, but all of those elements fabricated up to that moment were revisited by the welding team.

It is logical to believe that higher-strength steel should be used to carry larger loads and that low-strength steel should be used to carry smaller loads. We set out to dispel that logic by examining some trusswork near the base of the services core.

First, it is important to recognize that one uses high-strength steels to deal with high stresses, not necessarily to deal with high loads. Second, the structural engineer has the power to make loads go where he or she may want them to go; the desired path may not always be straight down.

Recalling also that most of the wind-induced overturning moment is carried in the triangular services core, it became obvious that we had to place as much weight on the corners of the services core as practical, minimizing the load on the columns in between the corners of the services core. We could not, then, allow the gravity loads to go straight down the columns.

How to solve this dilemma?

In order to move the gravity loads from the center of this system to the edge, we had to soften the system in the center, or stiffen the system at the edge, or both. We simply took advantage of the fact that, for the same load, higher-strength materials can be made more flexible than can lower-strength materials. When we do so, the loads in the center section go down, but the stresses go up. In short, by using high-strength materials in the center portion, we softened that section, forcing the gravity loads to move away from the center of the core, to the corner members.

By making use of nonconventional base plates and grillage details, we were able to significantly reduce the cost and the time of construction of the load transfer from columns to footings.

For the US Steel Tower, as in the contemporaneous World Trade Center, data was provided to steel fabricators in the form of IBM punched cards. To the best our knowledge, these two buildings were the first where data had been transmitted to steel fabricators in a digital format, allowing the shop drawings and some fabrication to be produced without human intervention.

As for the construction cost of the facade, the cost of construction for building services (mechanical, electrical, plumbing, and the like) generally run higher than the cost of the structural system. It is good to keep in mind that the coordination between the structural systems and the mechanical systems is an eternal conflict. The situation is not helped by the fact that, sometimes, the mechanical engineer delays his or her work until the structural system is firmly developed, leading to expensive changes for the structural engineer. While not to my knowledge, the reverse may be true.

I believe the architectural design of the US Steel Tower fails to rise to greatness. I attribute this to two main factors. First, I believe that there was too much input from me, the structural engineer, without a corresponding level of philosophic stalwartness on the part of the architects.

And secondly, the great "openness" of the megastructure concept is not reflected in the facade design. I attribute this to the insistence of US Steel to use their proprietary facade system and, perhaps, the architect's unwillingness to do battle. In my view, as I argued at the time, an all-glass facade between each of the giant spandrels of the megastructure would have brought tremendous power—greatness—to the architectural design.

Even so, the "Rusty Triangle" is a familiar landmark, admired and well known by all of the citizens of Pittsburgh.

At the moment of this writing (2016), there is a movement afoot to add public spaces atop the building. Called High Point, the new facilities would include an observation platform, museum, restaurants, educational facilities, and the like. Since we had designed the roof for vertical takeoff jet aircraft, there is little doubt that the structural system would safely support this additional load. The real issues, of course, are the required stairs and elevators. I have proposed a change from single-deck to double-deck elevator cars, rising only to the upper areas of the building. Above that point, stairs and elevators located in the re-entrant triangular spaces would carry, past the mechanical spaces, to the roof.

In closing, I share a thought of long gestation which I believe worthy of consideration: When the US Steel Tower opened in 1970, there was a fabulous party at the top of the building. Dignitaries from far and wide were present, including all of those most involved in the development, the design, the construction, the financing, and the leasing of the project, all of whom were properly introduced. The lovely and very talented woman who designed the mezzanine furniture was introduced, but the name of our company was not. To this day, that omission has circled in my inner being, overcome only by the satisfaction of having accomplished very significant improvements in so many facets of the design and construction of the US Steel Tower and in bringing it into the urban landscape.

Wasn't it Frank Lloyd Wright who, when asked how to improve Pittsburgh, curtly replied "Abandon it"? It seems to me that, instead of turning away from all that was so wonderful about Pittsburgh, the good citizens elected to combine the old with the new so as to create an extraordinary city. We are proud to have made our own contributions to that process, even where not acknowledged.

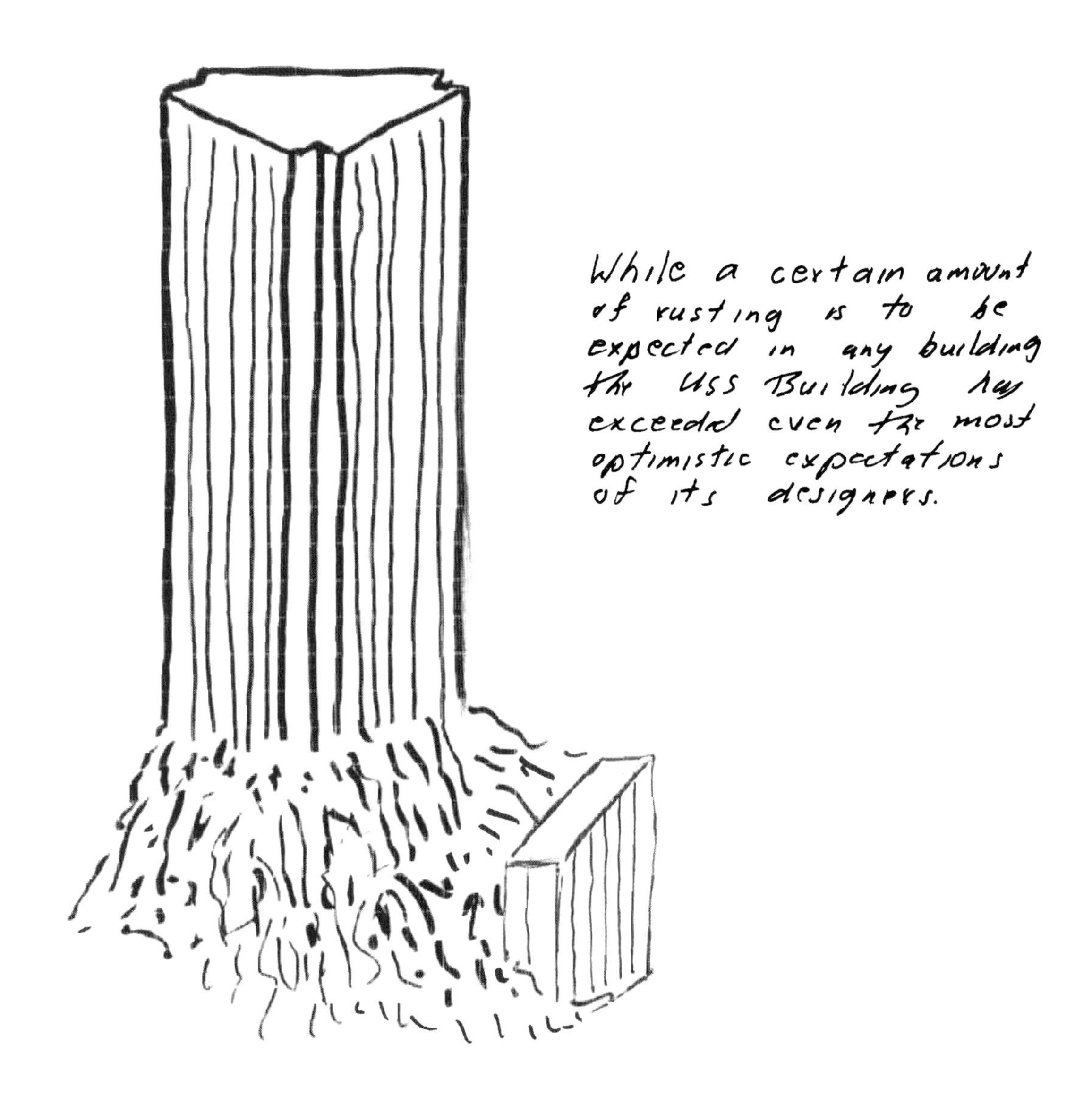
While a certain amount
of rusting is to be
expected in any building
the USS Building has
exceeded even the most
optimistic expectations
of its designers.

Gunnar Birkerts

1925 –

It is difficult to describe this architect, this Gunnar Birkerts. The physical part is easy: while now, 2016, physically a bit hesitant, he is dignified, handsome, tall, and thin, with longish hair. Being Latvian by birth, and having lived in Germany and the United States, his accent is unusual but comes softly to the ears.

By way of a little biographical background, Gunnar and his lovely wife, Sylvia (together in the photograph opposite), to the best of my knowledge, grew up and met in Riga, Latvia. Toward the end of World War II, prompted by the advance of the Russian army, they moved first to study at the university in Stuttgart. Thereafter they moved to the United States, where Gunnar worked with Perkins+Will and then Eero Saarinen before joining Minoru Yamasaki, becoming a principal and chief designer in the firm. It was 1996 when Gunnar set up his own practice to the north of metropolitan Detroit.

My introduction to Gunnar came when, without an appointment, he arrived at our offices in New York. It was 1967. My phone rang with our receptionist at the front desk saying, "There is an architect, a Mr. Birkerts, to see you." I almost instructed her to send him in but, instead, I rushed to our entry and personally ushered Gunnar into my office. We exchanged a few pleasantries and then he said, "Of course, you know my work." Not one to sugar-coat words, I told him the truth: "Gunnar, I've never heard of you before in my life." While, at first, he was taken aback, my response established a relationship between us that he has never forgotten. Gunnar understood then and understands now, that I'm always up-front in my communications with him and all design professionals.

My first project with Gunnar was the wonderful Federal Reserve Building in Minneapolis. He would later come to me with imaginative designs for other projects, large and small, about a dozen in all, including the Corning Fire Station (1974), an office building project for the US Embassy in Helsinki, Finland (1975), the Corning Museum of Glass (1980), and the US Embassy in Caracas, Venezuela (1992).

His most recent building, the magnificent National Library of Riga, Latvia, opened in 2014, twenty-five years after Gunnar first received the commission. As a native son, Gunnar's design was deeply woven into his inner being. The organization of the project, including the international composition of the extended design team, was very complex, though Gunnar, ever patient, sailed through the chaos with aplomb. Our role was to complete the structural documentation through Design Development but, realistically, the manager, squeezed out of us the detailing of sunscreens and other nonstructural elements. Because of the angled roofs, with pedestrians below, it was essential to control the sliding of snow. We proposed the use of occasional slots through the sloped roof, allowing the sliding snow and ice to fall into long pits, where it was melted and drained away. The experimental work was accomplished at the laboratory of Rowan Williams Davies & Irwin Inc. (RWDI), Toronto.

Thinking back over Gunnar's long career, it is clear that he has not been swayed by the architectural "fads" of the era. Straight and true throughout, his buildings have remained philosophically consistent. There is an unmistakable continuity in the beauty and utility of his designs.

As well, because of his long attention span in dealing with folks like me, and the small size of his staff, it has always been easy to communicate with Gunnar. I have learned much from each of our meetings, with the learning always in least expected areas. Some measure of our exchanges can perhaps be gotten from a letter I wrote in 1999.

After nearly a half century, Gunnar and Sylvia remain our good friends. Over Christmas, SawTeen and I sometimes traveled to the Birkerts' home to celebrate the year behind and the year ahead. While we never met their son, Sven, we read with care and admiration the essay he wrote on his illustrious father in "Gunnar Birkerts, Metaphoric Modernist."

It has been a privilege and an honor to work with such a creative architect and true gentleman—and one who has not been appropriately recognized by other architects of his time, or by the press.

With Gunnar meandering through the frigid streets of Riga, Latvia in 2007.

Dear Gunnar and Sylvia:

When I think of the two of you the mind turns to the subject of ideas. While flying from JFK to Narita, I found a really nice idea. It was one of those quiet ideas, a bit like an ember lying dormant in the ash alive and seeking to be found: waiting for me to dig it out before fading from its glowing self into grayness and oblivion.

Some ideas, these children of the brain, are so hot that one can't allow them to rest to scald the inner self. Such ideas need be tossed around in the mind, keeping them from arriving at any one shape. Then, late in the tossing, they become somehow molded into new forms into more holistic formats. In a sense, some ideas are a bit like the 12-bar structure of the blues; there seems to be an infinite combination of notes that can be taken from the one structure.

Gunnar once told me that ideas float around in the cosmos, waiting for someone to reach up, to pick them out of the sky. I prefer to think that my ideas cannot be plucked, that they can be only replicated. For sure, everything that I say or write, every idea that I take from the heavens, is constructed atop the shoulders of other mortals. When I think on the subject of "ideas" I realize that I take in everything that I can find and that I allow in as much accident as possible. I'm always envious of Gunnar, who seems to be so ordered in his thinking. For sure, with me, when I don't know what I'm doing, it's sometimes amazing what I get! It seems wonderful, that which I can sometimes find in emptiness. It's nice to find ideas. I like that a lot. Sometimes I worry an idea too much hoping that, through the process, the idea will somehow become better. It's good to recall that, when you've reached bottom, you should stop digging!

That thought brings to mind the story of a missionary couple who, with the help of native porters, was traveling on foot to their first post in a remote region. After several miles, the bearers took the bags from their shoulders and retired into the shade. A half-hour later, they remained resting, bringing concern to the minds of the missionaries; unless they hurried, they would not reach their destination by nightfall. "Shouldn't we move on?" queried the missionaries. "Not yet," responded the chief guide, "The porters fear that they've traveled too far, too fast. Now they're waiting until their souls can catch up and reunite with their bodies."

In any event, I sense often that my soul is racing to catch up with my mind... and that I should take a seat in the shade. Often, walking at God's speed, or just sitting in the shade, it is possible for me to enjoy the tension, the unexplainibility and the excitement of life. In the pre-dawn of the morning, I like to go out, adjust the horizon...and to then start my day.

Les

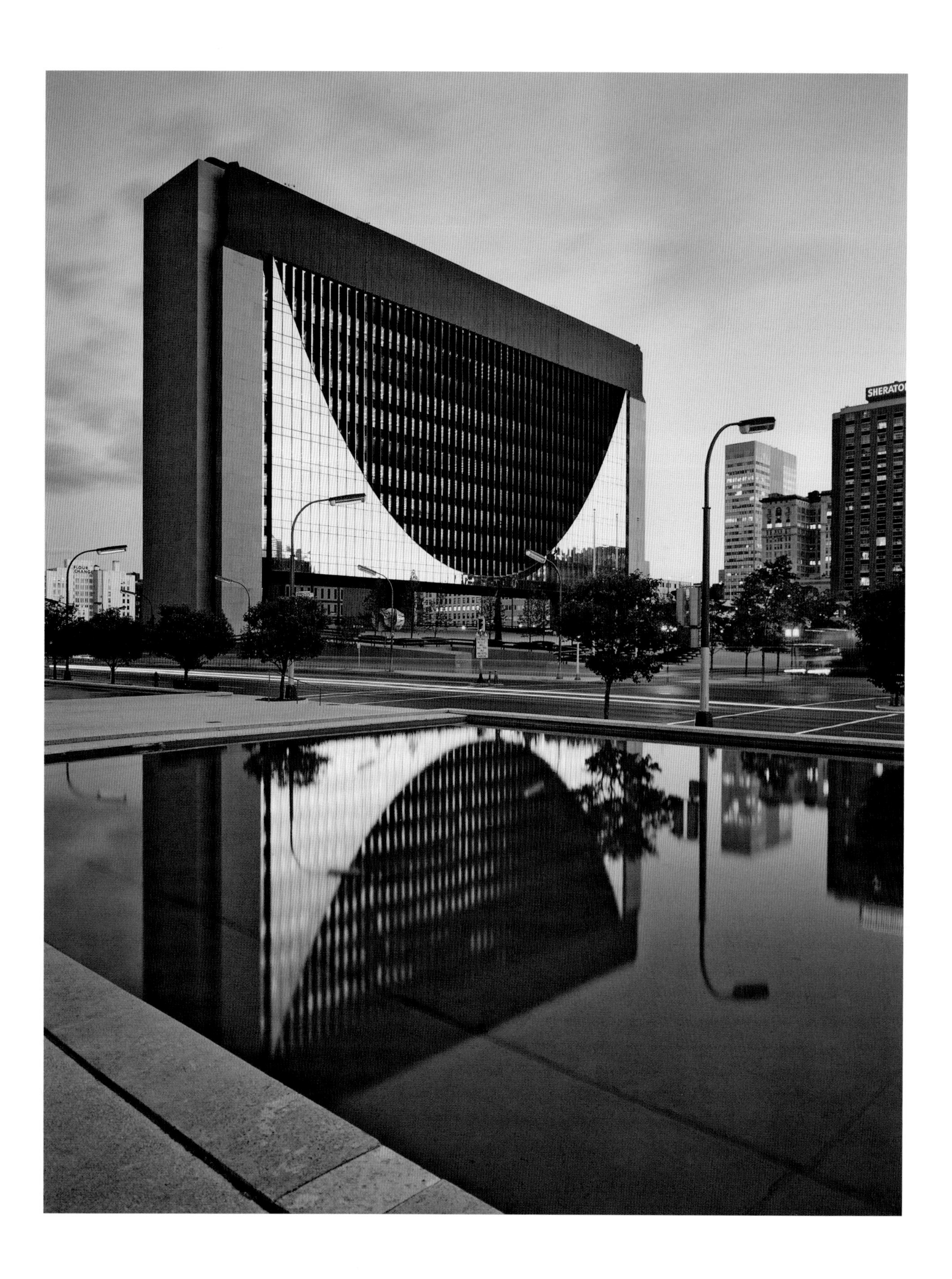
SHERATO

Federal Reserve Bank

MINNEAPOLIS, MINNESOTA. 1968

In opening this subject, it should be noted that it was the imaginative Jack Christiansen—we were once partners and remain friends—who accomplished the vast majority of the designs for this incredible project.

At our first meeting, Gunnar showed me some drawings of his proposal for the Federal Reserve Bank of Minneapolis (now Marquette Plaza): a building on two supports, 330 feet apart, with a column-free open plaza below. The breathtaking design provided an exciting challenge that begged to be met. He asked if the span could be achieved. My response was simple: "There isn't the slightest doubt that your long-span building is completely inside the technologies of both design and construction. Further, a variety of structural solutions should be explored in order to provide you with the solution that best responds to the architectural concept." He nodded approvingly.

The only significant difference between the building depicted in Gunnar's drawings and the as-constructed building is that the original drawings showed multiple catenaries supporting the clear-span building. I offered him two alternatives: a single catenary on each face to replace the multiple catenaries, explaining that the single catenary provided the vast majority of the support for the building, with the other, shallower catenaries, being more decorative than functional, and alternatively a single catenary down the center of the building, spanning the 330-feet between towers, while omitting the catenaries on each face.

Using simple sketches, I was able to convince Gunnar that just the one catenary, the deepest one, was quite sufficient for the safe and the economic support of the building. I went on to point out that the other catenaries were useful for the circumstance where the deepest catenary failed but that, with proper design, the single catenary could provide an adequate level of robustness and redundancy. The words *redundancy* and *robustness* were new to his ears but, I suspect, it was those concepts

The structural system was fully expressed during construction. Alas, when the building was sold years later, the new owners filled in the space below the tower with retail, completely destroying the sense of the clear span.

deep in his mind that had led him to the multiple-catenary approach.

Another design consideration that we discussed was the client requirement for the future addition of 50 percent more floor area. Gunnar's solution was to add the expansion directly above, all carried by arches mirroring the catenaries below. Gunnar was concerned that the shallower addition could create problems with the building below. I pointed out that, since the "future" addition would not rely on the catenary below, his proposal actually reduced the axial forces carried by the stiffening truss housing the mechanical spaces at the top of the building. In our discussion Gunnar came to understand that the horizontal thrust of the shallower catenary above was more or less identical to the horizontal thrust from the deeper catenary below—the upper catenary being in compression and the lower catenary being in tension. We also discussed various ways of providing the additional elevators that would be required.

Why the clear span of the building? Gunnar pointed out that the below-grade areas were to be largely occupied by vaults, armored trucks, and 18-wheelers carrying money, papers, and other secure documents. Being column-free, the long-span structure greatly enhanced the usability of this below-grade area.

Gunnar's brilliant solution to the client's requirement for future expansion was to include the addition directly above.

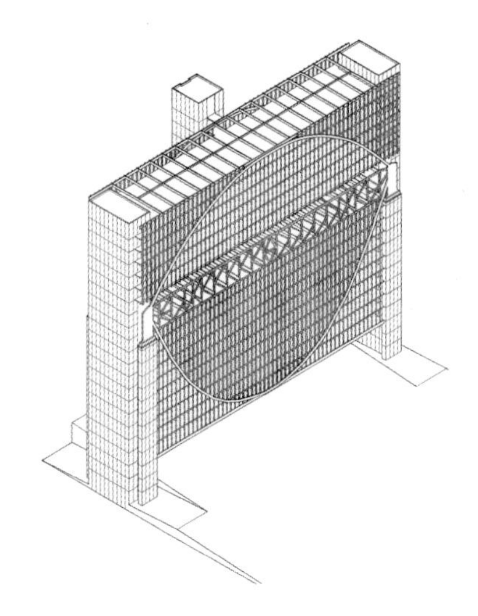

This first meeting with Gunnar was an interesting and exciting experience for me, an opportunity to exchange ideas with this wonderfully talented architect. While we discussed each of my proposals in juxtaposition to his drawings, Gunnar was not one to jump to conclusions. The decision was to wait for another day. That day came, but Alas! part way through the design I found myself completely swamped with other projects, most notably, the World Trade Center and the US Steel Tower in Pittsburgh. Losing all too much sleep in the worrying, I asked the very talented Jack Christiansen in our Seattle office if he would take the helm, to which he readily agreed. After consultation with Gunnar, Jack assumed control over the structural design.

Well after the completion of the project, a Vice President of the Federal Reserve System invited Gunnar and me to lunch, there to discuss "problems" with the building. He opened the discussion noting that "During the design of the building, the only thing that really concerned us was the structure. Now, however, the only thing that really works well is the structure."

With the passage of time, a series of problems had developed. The mechanical system centered about a radiant ceiling for heating and cooling. The fireproofing system for the steel floor beams and the steel columns was asbestos-based, requiring that changes and repairs to the mechanical/plumbing/electrical systems be accomplished by workers wearing masks and other protective garb. While, during the design, I had written to all who would listen the many reasons why asbestos should not be used (and I did not then know of the material's intrinsic health hazards), perhaps for issues of cost, the material was incorporated into the work. Short of stripping out the asbestos, there was little that could be done.

Because the catenary enclosures acted as chimneys, frigid winter air spilled inside, making untenable the adjacent work areas. Fortunately, an economical solution was developed.

A second problem was that, during the cold winter months, occupants refused to sit in areas adjacent to the catenary arches. Because of stack action, where the catenaries were not well sealed at the floor line, the catenary enclosures acted as chimneys, driving the frigid Minneapolis air into the occupied spaces and subjecting persons in the vicinity to bitterly cold drafts. Clearly, the between-floor insulation and/or fire-stopping was not up to the job. The matter was completely outside of the technical expertise of both Gunnar and me. Still, having been a pioneer in the issue of stack pressures, I stepped up to the plate, volunteering to resolve the matter. I developed a concept to seal off the airflow and then had wiser heads than mine sort out the detailing.

A final big problem was that corrosion had appeared in the metal facade. Again, not being truly expert in such matters, I sat on my hands, and Gunnar addressed the rust with his curtain wall consultant.

Some years later, in 1997, the Federal Reserve sold the building to a developer. While we could do nothing but weep, said developer closed off the bottom of the building, changing it from an open column-free space into a walled-in retail space. This totally destroyed the appearance of the building as spanning freely between the two towers. Although there was some local protest, the architectural community failed to mount a successful campaign against this destruction of an important achievement in downtown Minneapolis, a true landmark in modern architecture. I did not become aware of the matter until it was a *fait accompli*. While attempting to rationalize such things, one can only conclude that there is no explaining this assault on Gunnar's designs.

SUMARÉ PLANT

INTERVENTION

IBM Manufacturing & Production Plant

SUMARÉ, BRAZIL. 1971

On prior occasions, on various endeavors, I had worked with John Novomeski Jr. of IBM Corporation. He called in 1971, asking me to stop by to discuss the IBM manufacturing facility in Sumaré, Federated Republic of Brazil; the facility was suffering from excessive deflections of the roof. It seems that an IBM real estate manager, while visiting the factory, observed that, in a rainstorm (of which there are many in southern Brazil), the roof gutters overflowed, spilling water onto the manufacturing operations below and had so reported to New York. John asked me to review the structural systems. I countered with the proposal that I first visit the site, examine the reported difficulties, interview the architects and the engineers as well as workers in the facility, and to then report my findings. This agreed, that very day, I took an overnight flight to Rio de Janeiro. I am told that John, relating to his term in the US Army and to my quick departure to Rio, had said: "You have to know where the decisive point is; that is where a leader is supposed to be."

IBM's local management, in Rio de Janeiro, seemed totally disinterested in the matter, contending there were no issues at the Sumaré plant that were not common to all such buildings. It was much the same at the architect's office, where I was told that no one in-house was familiar with the project, nor was any such person expected for an unknown length of time. It was abundantly clear that neither IBM-Brazil nor the architects were prepared to cooperate with me.

After two discouraging days in Rio, I took the first flight to São Paulo, rented a car, and drove the 120 kilometers northwest to Sumaré. Without credentials, but determined to achieve my goal, I bluffed my way into the IBM facility. Having no command of either Portuguese or Spanish, I found no one who could, or would, communicate with me. In desperation, I wandered along the hallways of the administrative areas looking for I knew not what. From behind the closed doors of a conference room, I heard a voice speaking a mixture of Spanish and Portuguese with some English words along the way. Suspecting that this was a person of some importance I barged into the conference room, asking if anyone spoke English. An IBM employee stepped to the fore, and my adventures began.

The two of us meandered through the plant. My eyes and ears discovered the following:

- The roof was supported by a space frame consisting of hot-rolled steel, angle-shaped members connected by T-shaped bolts;
- Columns were approximately 100 meters on center;
- The rain gutters, on a 20-meter-square span, were wide with open channels, supported by the structure of the roof;
- It was confirmed that, in a rainstorm, the weight of the water would deflect the roof, allowing rainwater to spill out of the gutters onto the manufacturing spaces below;
- The floors showed countless signs of water having fallen and subsequently evaporated.

Climbing onto the roof, I noted that the gutters were sensibly connected to the space-frame roof structure. Of some interest, frogs and tadpoles could be seen in lingering puddles, but there was no path from the ground to the roof. Did the tadpoles fall from the sky? The amphibian mystery went unresolved, but as I walked I could both see and feel the roof's vertical deflection. That roof was extraordinarily flexible!

Returning to the manufacturing floor, a quick calculation showed that the steel structure and the fiberglass roofing, plus the rain gutters, weighed about the same as 50-millimeter-thick wood planking. This was an extremely low weight for a roof on a 20-meter span!

I conducted a cursory inspection of the entire building, and after collecting the architectural and structural drawings I headed to the airport at Campinas for an overnight flight back to New York. Standing in a quiet corridor, away from the bustle of the main terminal, a young man came dashing around the corner, stopping a few feet away. In Spanish/English he explained that he was Pelé, Yes, the famous football (soccer) star—hiding from the press. We ended up sharing a bottle of Guaraná, that delicious, apple-flavored soft drink of Brazil.

Back in New York, I related my findings to IBM and sorted out a handshake agreement for the checking of the design and all else that might be required. Thus began a series of weekly trips: New York–Rio de Janeiro–São Paulo–Sumaré–Campinas–New York, departing Thursday night and returning Friday night. I soon became friends with the Pan Am crews and was allowed, in the era before "sleeper seats," to put an air-mattress on the floor so that I could get a bit of sleep.

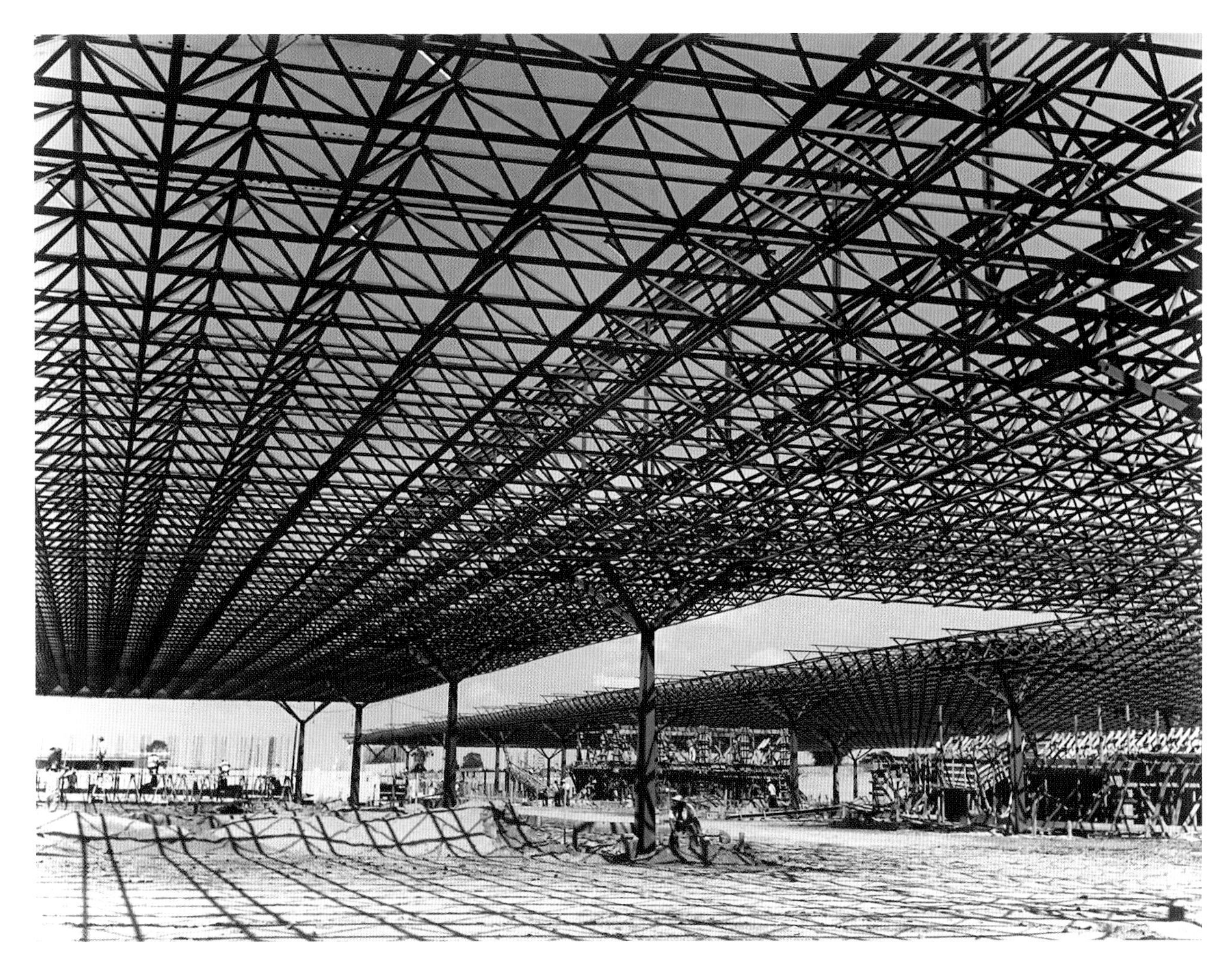

The nature of the structural system, coupled to the vast number of connections required for this space-frame structure, are apparent in this construction photograph.

There is a truism in structural engineering for planar structures (e.g., IBM's roof in Sumaré): The average of the end moments plus the centerline moments are equal to the uniform load times the span squared, divided by eight ($wL^2/8$). This truism led me to conclude that the strength of the roof structure was significantly below that required for an appropriate design.

Concerned about a possible collapse, I ordered a full-scale load test of a portion of the roof. In full realization that the deliberate increase of loading could lead to the progressive collapse of the entire roof, protective shoring was carefully designed and the test procedure planned to allow the immediate reduction of the test loads.

While preparations were being completed, just on a hunch, I had carried to New York for testing a small bundle of those unusual, T-shaped bolts used in the connections of the individual members of the space-frame. The bolts had been specified by the structural engineer to be high-tensile bolts but on our testing proved to be common bolts, with a strength of less than half of that assumed in the design! I made the discovery just in time to cancel the above-cited load test of the roof.

Above The structure's typical space-frame connection.

Top right The unusual bolts used by the original structural engineer proved to be of ordinary strength, not the high-strength bolts required.

Adding these two facts together ($wL^2/8$) and understrength bolts, I met with a few of IBM's executives, recommending that the building be evacuated. They requested, and received, a detailed summary of the defects that I had found. I was told that John Novomeski then met with high officials of IBM before putting aside all issues of cost and production schedules, ordering the evacuation.

An emergency meeting was called with IBM's top management in both New York and Brazil together with the Brazilian architects and the Canadian structural engineer, the latter teaching at a noted Canadian university. I reached the professor by phone to provide the time and the place of the meeting. He explained that he had classes and research projects that could not be delayed. I countered that we had evacuated all persons from a building of his design, that the meeting could not be delayed, that he must attend. It was not possible, he insisted, and abruptly our conversation terminated. I immediately telephoned the provost of his university to explain the situation, demanding the professor's participation. Almost immediately, the provost called back: the professor would attend the meeting.

Concerned with legal consequences, I did not provide the professor with the details of our findings prior to the meeting, stating only that we may have uncovered serious issues regarding the design and the construction of his project.

The meeting opened with an introduction of all the attendees, and then IBM turned the proceedings over to me. I elected to ease the high tension in the room with an overview of my initial conversation with John Novomeski and my subsequent travels to São Paulo and Sumaré. The professor interrupted, saying that he had not come to listen to such "trivia."

With this almost perfect introduction, I spoke first of the $wL^2/8$ problem and of the load test we had designed. I then addressed the matter of the "high-strength" bolts that proved to be under-strength, resulting in the cancellation of the load test. At this point, an IBM executive asked the simple question: "What does this mean?"

There was total silence as all eyes focused on the professor. The silence continued as he, the designer of the structure, looked down at the floor. Everyone would have been delighted to hear that the structure was safe, that they need deal only with the excessive deflections of the roof structure.

After an interminable moment, the professor looked up to say, and I quote with precision: "You got me." The silence was deafening.

Someone, I cannot now recall who, asked the professor if he had anything more to say, receiving in response only a head-shake. The professor then left. While there was little more to say, I responded to a question from IBM by stating that, in my view, the structure was significantly under-designed, and that an event leading to collapse in one area of the roof could lead to the progressive collapse of the entire manufacturing facility. I suggested that another engineer be retained to evaluate our findings.

IBM ordered the facility to remain evacuated and, neglecting my advice, retained us to develop the required repairs and to provide an on site engineer to evaluate work in the field. We sent our very experienced and competent Jim White, who had led the surveillance activities of our field-engineers for the World Trade Center in New York. My own participation for the IBM plant was limited to the analysis and the development of repair details and continuing Friday site visits to Sumaré. Approximately 50,000 repairs were completed under Jim's watchful eye.

Regarding the architects, the engineer, and the contractor, closing my eyes and ears to such things, I am not privy to the legal actions on the part of IBM, if any, that ensued. Indicating something of the professionalism that we had displayed, when all was completed, the contractor asked for our assistance on another project.

I cannot complete this discussion without iterating that, in my view, IBM's executives displayed a totally professional and humanistic attitude toward these events. Putting aside economic consequences in favor of protecting the work force in the plant, they had significantly raised my appreciation of the importance of thoughtful corporate responsibility. I can only hope that these qualities are prevalent today and will so continue into the future.

Mitchell/Giurgola Architects

One late afternoon in the 1970s, I took a seat on a bench at the Museum of Modern Art in New York, there to admire a painting. Shortly, on the same bench, a man took a seat next to me. We may have exchanged a word or two before he departed. I stayed for another ten minutes or so before heading home.

On the following day I had an appointment with Mitchell/Giurgola Architects at their New York offices on the Upper West Side of Manhattan. Much to our mutual amazement, Romaldo Giurgola and the man at the Museum of Modern Art were the same person. What a fortuitous and wonderful introduction to a new client! While the above was a matter of pure happenstance, a young architect or engineer, by attending museum and gallery openings, is able to meet all sorts of creative people—but only where there is a genuine love of the arts.

After chatting about many things, it was agreed that we would work together on a new Law Library at the University of Washington in Seattle. Some twenty wonderful buildings were to follow.

With offices in Philadelphia and New York, Ehrman B. (Mitch) Mitchell and Romaldo (Aldo) Giurgola established their firm in 1958, enjoying commissions from a wide variety of clients. While these architects were very different in personality, their two offices had a commonality that, at least to me, was surprising. Both were delightful people, and a pleasure to work with.

Mitch, born in Pennsylvania, earned a Bachelor of Architecture, summa cum laude from Pennsylvania State University. He acted very much as the organizer of the firm, keeping projects on track and on budget, but was a fine designer in his own right. We were very appreciative of his input into those projects assigned to us. While overshadowed as a designer by Aldo Giurgola, one need only admire his own house to understand that he was an architect of many talents.

Born in Rome in 1920, Romaldo Giurgola was educated in Italy before travelling to the United States to study at Columbia University, later to lead the New York offices. Beyond Aldo's talents as an architectural designer, he was blessed with the wonderful ability to draw and to sketch. His perspective drawings are nothing less than breathtaking.

One afternoon, in a moment of insanity, I suggested to Aldo that we could block out an outline that could be used as the basis of his perspective, thus speeding his work. Smiling, he said: "Well, these things are more difficult than your computer may imagine. But, if you wish, try it." Of course, we did exactly that, printing our outline on translucent tracing paper, believing that it could be used as the basis for Aldo's own perspective drawing. In Aldo's absence, I laid our trace over his drawing, there to discover that, in subltle ways, he had modified the true perspective to make the final rendering even more exciting, more beautiful. Realizing that our computer was able to produce a true perspective, but not a nuanced one, I rolled up our drawing and crept away. Yes, we continue to learn. Of course, since that time, the firm has become computer-savvy. But I cannot believe that any computer has risen to Aldo's level of excellence.

The Mitchell/Giurgola project that brought us the most personal friends is the San Jose Convention Center, just south of San Francisco. While the brunt of the work was carried out in Philadelphia, Aldo participated heavily in the design, thus further solidifying our relationship. Lauren Mallas and Fred Foote of Philadelphia were the team leaders but it was Robert (Bob) Schuman who directed the detailing of the architectural work and who, working with Aldo, was most active and most appreciated. Indeed, we remain close friends with Bob, his architect-wife Joyce Lenhardt and their family. SawTeen, carrying the brunt of our effort, was our Partner-In-Charge.

With Aldo's permanent departure to Australia for the Parliament House project in Canberra the firm split into three separate offices: Mitchell/Giurgola Architects, LLP in New York, MGA Partners Architects in Philadelphia, and Mitchell/Giurgola &Thorp Architects, in Australia.

It is a long way between New York and Canberra, a trip made more difficult by our respective heavy workloads. Still, we remained in touch with both Aldo and Harold (Hal) Guida, who had moved to Canberra with Aldo, and stayed on. Many years later, in 2010, while on our way to Melbourne, SawTeen and I stopped in to visit with Aldo, who was frail of health but sound of mind. His wife, the delightful Adelaide, had passed a decade earlier but their daughter, Paola, lived nearby. While our visit was focused on Aldo, we did take the time to visit Parliament House in Hal's company there to enjoy many of Aldo's drawings and renderings in an exhibition celebrating his ninetieth birthday. Aldo was happy, as were we, that SawTeen and I had taken time from our overly busy schedules to spend such delightful hours together.

While not obvious from this construction photo, the slope of the bowstring trusses is constant regardless of the changing spans. The vertically cantilevered concrete work and housing services provide the basic earthquake resistance for the roof and for the concrete work below.

Opposite Aldo Giurgola's early sketch for Parliament House and its iconic Flag Mast.

Parliament House

CANBERRA, AUSTRALIA. 1979

In 1979, Aldo was invited to join the panel of judges for an international design competition for the Australian Parliament House, in Canberra; rather than judge, he elected to compete and invited me to join his team. There were 329 entrants into the architectural competition for the design, with many of the designers being both exceptionally talented and internationally acclaimed. Of course, nearly all of the more noted architectural teams in Australia were included, as were over 130 wonderful architects from 27 countries.

We had a key player, Richard (Rick) Graham Thorp, who was licensed to practice in Australia and was temporarily employed in the New York offices of Mitchell/Giurgola Associates. While Aldo was clearly the team leader and the principal designer, Rick brought in-depth experience of the nuances of Australia plus a tireless determination to win the competition. He worked endless hours in designing, organizing, and accomplishing the competition documents, and drove me to produce complimentary documents emphasizing the compatibility of the structural systems with the architecture; of course, all of this was done without financial compensation.

Both Aldo and Rick were open to my ideas for the betterment of both the architectural concept and the competition documents. However, it was Rick who provided the creativity and the energy that molded Aldo's ideas into the reality of the competition documentation. Of course, Aldo unleashed his incredible skills with pen and pencil, which drawings were key to the overall presentation. As each of Aldo's compelling drawings and sketches came into existence, the more detailed drawings were constantly revised to conform. One of my more important contributions was in the furthering of the development of the concept for the flag-mast structure. While Aldo had developed an imaginative concept, some significant modifications, some now long forgotten, were appropriate.

The jury consisted of high-level persons from a variety of relevant walks of life. My good friend I. M. Pei, one of the two architects on the jury, said: "Les, it was clear from the submittals that yours was the best entry, the only entry with the stature worthy of the project." Of course, he spoke from an architectural perspective whereas the jury as a whole needed to examine a host of other equally important variables.

Hey! We won! Better said, Aldo won!

After the competition winner was selected, Aldo, Rick, Hal Guida, and a few others moved to Australia, continuing to practice there.

MOVING FORWARD FROM THE COMPETITION

At that time, professional fees in Australia were more-or-less defined by the nature of the project. The hilltop site for the Parliament House was difficult, and reflections on our entry led to the need for changes. While holding firmly to Aldo's basic concept, in order to deal with the seemingly unending issues, where required in order to cement our concept design into a sense of reality, we made several trips to Australia. These were quiet times, leaving room for the enjoyment of travel into other reaches of Australia, a circumstance that led to a better understanding of these Aussies, these wonderful men and women.

On one flight, before continuing on to Sydney, my "nonstop" flight made an unscheduled landing in Auckland, New Zealand. Foolishly, not informed by Pan Am, I had assumed that the unscheduled stop was a matter of headwinds and fuel consumption. Alas, upon landing at Sydney, I discovered that almost all of the trades were responding to a labor strike. There were no rental cars available, there was no bus or rail transportation into Sydney, much less Canberra 180 miles further south. What to do? Sitting on my bag in the great entrance area of the airport, my mind groped with the problem. A man passed by with a "Hi, mate!" before continuing on his way.

But wait! I spotted a rental-car contract in his hands. I almost tackled the good man! It turned out that he was an architect, headed by air for some place to the north. I negotiated a deal whereby I paid for all of his rental car charges, he gave me his contract and the car keys, and I was blessed with his rental car. Of course, it was completely illegal, but I'd flown 10,000 miles from New York and was faced with the alternative of returning without completing my site evaluations.

Well, that car was put to good use. Following my work activities in Canberra, I drove to the east coast of Australia, overnighting here and there, but reaching as far south as Wilsons Promontory, Squeaky Beach, and the like. It was a solo journey, but great fun. As for Australian cuisine, I will confess that I failed to learn to enjoy Vegemite, and that spaghetti sandwiches were not for me, but from a good restaurant, one could obtain a fine meal and excellent Australian wines. Perhaps the food would have been better were it not for the fact that the beer truck drivers were out on strike. The restaurants were dry, dry, dry.

Reviewing the site plan while on Capital Hill, prior to construction.

Following my work and the fun-tour in Australia, I returned to the US and the disturbing news that the Parliament House Construction Authority had decided that the design effort and our offices must be located in Canberra, not Sydney where the population was larger, professionals were more diversified, and the city was a more exciting place to live and work. Earlier, for the World Trade Center, from the experience of having opened an office in New York, while my needs were for steady, imaginative, talented, forward-thinking structural engineers, I knew well that many of those applying for a position in Canberra would be "drifters," going from job to job, always seeking a higher salary, thinking of an office in Canberra as a one-project office. In my mind, the office should have been permanent, seeking work beyond the Parliament House project. Discussions with the Authority did not soften their position; the office had to be in Canberra.

With the most reluctant goal of resigning from the project, I set out to interview many of the structural engineering firms in eastern Australia. Meeting all of these talented engineers was an informative experience. The standard response was "We will create an office in Canberra, staffing it with qualified engineers." While that all sounded fine, none seemed prepared to provide the name of that fine designer who would lead the team. And then I met John Fowler of Irwin Johnston & Partners. John, a talented and personable engineer, fit right onto the Parliament House team, saying: "I'll move my family to Canberra, set up a group from my offices in Melbourne, plus recruit some additional staff in Canberra." As well, John and his wife, Jeanette, were the sort who would complement Aldo and his team. It was a perfect union.

In essence, in favor of John Fowler, we resigned from the project. Well, John did ask me to participate in the structural design for the Flag Mast. More importantly, John, Jeanette, and I remain good friends to this day. It was just a few years ago, over a fine dinner in Melbourne, when SawTeen and I met last with John.

Citicorp Center

NEW YORK, NEW YORK. 1978

It was the dawning of August 1978 when, following breakfast and preparing to depart for our offices, the phone rang. An unidentified lawyer requested that I join "a meeting." When I asked about the agenda, he told me only that it was "important." The conversation was curt: "What's this all about?" I inquired. "You will find out at the meeting." After turning the matter over in my mind, but only for a few seconds, I responded: "Sorry, I don't operate on that basis."

The next morning, a second lawyer called with the same conversation, and the same result. A few moments later a third, more forthcoming lawyer called to say that there was a possible problem with the structure of Citicorp Center. It being not sensible to quibble with a *big* bank, I attended the meeting that very morning. It was the beginning of a period of mostly sleepless nights and high stress levels.

At the meeting, the only person I knew was William J. (Bill) LeMessurier (who passed in 2007), the structural engineer for the 59-story Citicorp Center. I gathered that most or all of the other attendees were lawyers or indemnity insurance carriers apparently representing the LeMessurier firm.

LeMessurier outlined the following problems:

While the connections of the building's major structural diagonals were constructed to carry the loads given in the structural drawings, the contractor had requested a change to less costly bolted connections instead of the welded joints originally specified. This change, accepted by the structural engineer, reduced the available tensile capacity of the connections while still retaining more than adequate capacity to carry the loads stipulated in the original structural drawings

Having been queried by a Princeton civil engineering student (I later learned that it was Ms. Diane Hartley), LeMessurier determined, making use of new wind loads developed by Professor Alan G. Davenport, that the original connection loads given in the structural drawings were not sufficiently strong to safely sustain the wind loads arriving from a diagonal direction, wind loads that were not considered in the original design but were proven to be critical to the integrity of the structure.

All heads then turned to me: "Well?" My response was short: "If this is in fact the case, you have a very serious problem." I went on to say that I would need much more information in order to provide any reliable assessment of the problem. The discussion continued, focusing largely on the need to communicate with the architect, Hugh Stubbins, and with Citicorp. Along the way one of the lawyers asked if I would assist in the defense of LeMessurier, to which I replied in the affirmative. Of some interest, I never met Mr. Stubbins (who passed in 2006).

After the meeting, I went to see Stanley Goldstein, a fine structural engineer who headed LeMessurier's office in New York, there to accomplish a cursory review of the extent of the structural drawings.

In the days following, in the LeMessurier New York offices, I spent some hours studying the drawings of Citicorp Center. And again, to my considerable surprise, in just studying the drawings, without initiating mathematical analyses, I uncovered yet another series of defects in the detailing of the structural system. These defects were serious, most notably, the lack of adequate lateral bracing of some of the bottom flanges of large bracing members of the perimeter framing and the inadequate torsional bracing of some of the columns in the services core. To the best of my knowledge, as of this writing, the existence of these defects, which I understood were fully repaired, have never been reported to anyone other than to LeMessurier's team, to representatives of Citicorp, and to the contractor.

I have read that LeMessurier considered various options: bury the matter, commit suicide, confess. "I had information that nobody else in the world had," he told Joe Morgenstern for his article "The Fifty-Nine-Story Crisis" in the *New Yorker* (May 29, 1995). "I had the power to initiate extraordinary events that only I could initiate. I mean, sixteen years to failure . . . that was very simple, very clear cut. I almost said, 'Thank you, dear Lord, for making this problem so sharply defined that there's no choice to make.'"

After a few days, Hans Angermueller, who appeared to head Citicorp's legal efforts, called to request a meeting. I responded that, having agreed to assist the LeMessurier legal team, I first needed to speak to them; he responded that they had already been contacted and that it had been agreed that I was to represent Citicorp, not LeMessurier. While it can be argued that I should have been consulted in the making of this decision, I was far too pleased to contest the matter.

The meeting with Angermueller and other Citicorp executives, perhaps others, brusque and to the point, took place in an intimidating, huge, book-lined conference room in their executive offices on Park Avenue. They wanted to know what would happen should they not undertake the proposed repairs. After first explaining that, not having examined the shop drawings and having completed only an initial review of the design drawings, there was much work to be done. I went on to state that significant further study was required. "Should the information in my hands be correct," I said, "within your lifetime, where essential repairs are not implemented, it is possible that there could be a wind-induced collapse of Citicorp Center."

"What does that mean?" they asked. Taking one of the large legal books from the surrounding shelves, and standing it upright on its narrow edge, I said: "The worst scenario would be—" and I toppled the large book to the table, which resulted in a resounding BANG! Indeed, I was surprised and startled by the intensity of the sound, but my demonstration had the desired effect. There would be no second-guessing of the need for repairs.

For reasons not fully understood by me, instead of making use of their offices in New York City, LeMessurier rented some space immediately contiguous to our offices. This was very convenient for me as I could walk just a few steps from my normal responsibilities to the center of the Citicorp analysis team. The engineers provided by LeMessurier, taken from their offices in Massachusetts, were highly skilled and eager to accomplish repairs that were both safe and practical.

While there were countless meetings, I recall one in particular attended by LeMessurier and some of his associates, together with Citicorp Vice Presidents Robert Dexter and Henry DeFord. LeMessurier outlined the proposal for the repairs, which proposal was tentatively accepted by all parties. Since Bethlehem Steel, the original contractor, was no longer in the business of fabricating and erecting structural steel, I suggested that Karl Koch, steel erector for the World Trade Center, be considered for the structural repairs. Looking for an immediate solution, neither Dexter nor DeFord wanted to "consider" alternatives—they wanted immediate action.

I got on the phone to Karl Koch only to hear: "We're a little busy right now." I then explained that we were discussing Citicorp Center; Karl's response was instantaneous. A couple of hours later, two of his engineers arrived at our meeting. LeMessurier and I took them to an area of an unoccupied floor where the sheetrock had been removed. Comparing the drawings with the exposed bolted connections, the two engineers agreed that the proposed repair was feasible. As it turned out, to my surprise, Koch had almost all of the required steel on hand. We were off and running!

I contacted a California firm to install strain gauges on the steel work in order to monitor the repair-related change in stress of selected members, associated for the most part with the wind. There were two thorny problems, the first being that the Electrical Workers Union objected to the use of non-union technicians for the

A diagram depicting the major connections of the perimeter framing in need of repair.

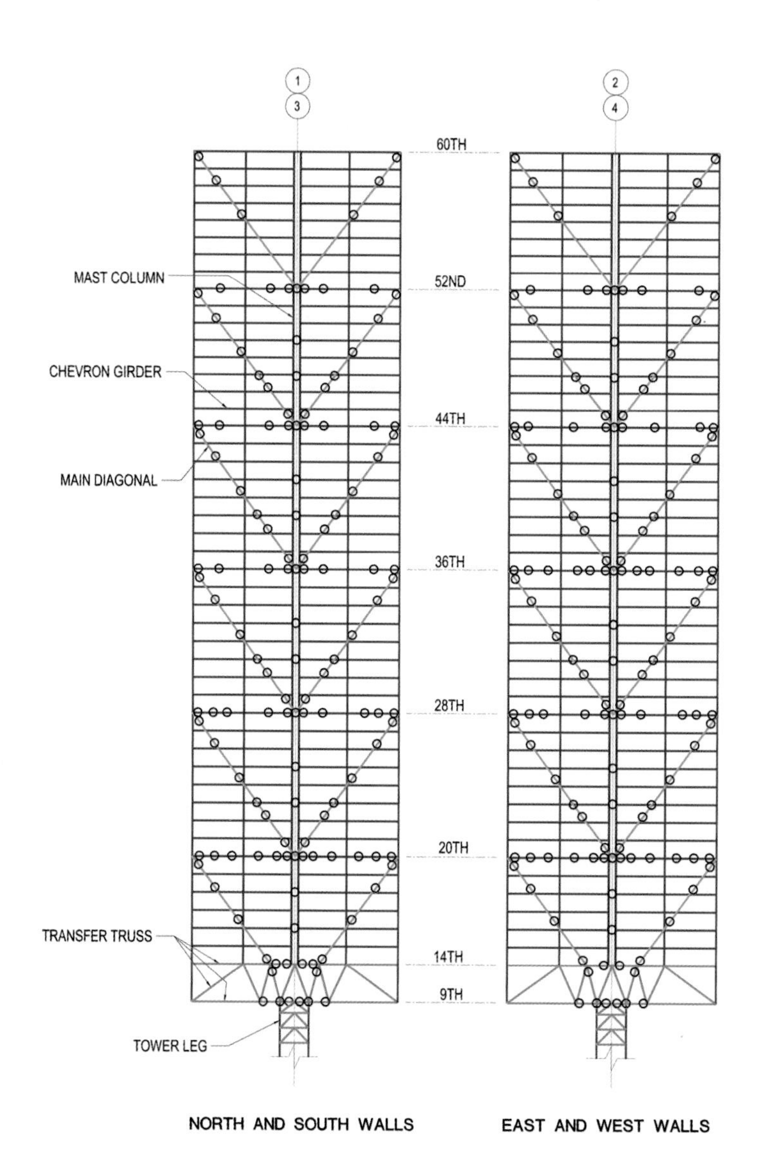

installation of the strain gauges. I sat down with a union representative to explain the nature of the electrical work. He listened thoughtfully but without knowledge of the dangers associated with the defective designs. After lengthy discussion, he acquiesced. The work went ahead, as I had planned, making use of the skilled non-union technicians.

Walter Wriston, Citicorp Chairman and CEO, telephoned, suggesting that he come to our offices seeking an in-depth, outside opinion of the Citicorp structural issues. Amazed at his request, I suggested that I come to his offices instead, which he accepted. This resulted in a more-or-less regular weekly meeting in Citicorp's building on Park Avenue.

Another problem was that we needed a hard-wired telephone line from these new strain gauges to recording equipment in our offices some blocks away. I was told that the line could be had in a few weeks' time—but I wanted that line immediately! The solution came in my next regular weekly meeting with Walter Wriston and President John S. Reed. Informed of the problem, Wriston called his friend Charles L. Brown, president and CEO of AT&T. Of interest to me, "Charlie" Brown answered the call personally, without secretarial intervention, indicating something of the ability of these high-ranking executives to communicate together. While nothing was said about the problems facing Citicorp, we had our phone line the next morning.

Further, to obtain expert advice, I contacted the US government's Brookhaven National Laboratory on Long Island to obtain the services of two scientists known to me: windspeed/weather experts to advise us several times each day on possible upcoming high winds. Citicorp readily accepted this arrangement.

Jim White of our offices contacted structural steel inspectors from New York and New Jersey to find truly expert welding inspectors who would work the midnight shift; Jim's success was truly outstanding.

On August 8th we hosted a meeting in our offices with the Commissioner of the New York City's Building Department and a small group of senior officials seemingly from all of the Boroughs of New York. Realizing that the session carried high, potentially political, and economic consequences, I asked that I attend, but sitting in the back row of seats, providing advice only where needed. Our phones rang throughout the meeting: newspaper reporters wanting to know what was going on, why the meeting was being held in our offices, and why they had not been invited. Fortuitously, the meeting coincided with the beginning of a newspaper labor strike, which continued throughout the period of the repairs. With the temporary suspension of all New York newspapers into November, our work went quietly ahead. I have read, but find it difficult to believe, that Bill LeMessurier had persuaded Citicorp to repair the building without informing the public, a task made easier by the newspaper strike. Indeed, the meeting with the Commissioner of the New York City's Building Department seemed to me to be an "official" informing of the public, which eased my concerns regarding our own ethical behavior.

During my late Wednesday afternoon meetings with Mr. Wriston, usually with Mr. Reed present, all other persons were barred from the floor; the secretarial and administrative staff departed when I arrived. In one meeting early on, having worked with the National Science Foundation and the National Research Council, and having served as vice president of the Wind Engineering Research Council, I tabled something of my own experiences in disaster management. With this as background, I provided my contacts at various disaster-relief organizations, leaving open the question of whether or not such contacts should be initiated. Later, Bob Dexter and Henry DeFord, following my advice, coordinated with the Director of Disaster Relief for metropolitan New York at the American Red Cross.

The storm track of Hurricane Ella as it headed for the United States before turning to the north-east and, with little damage, passing well south of New York City.

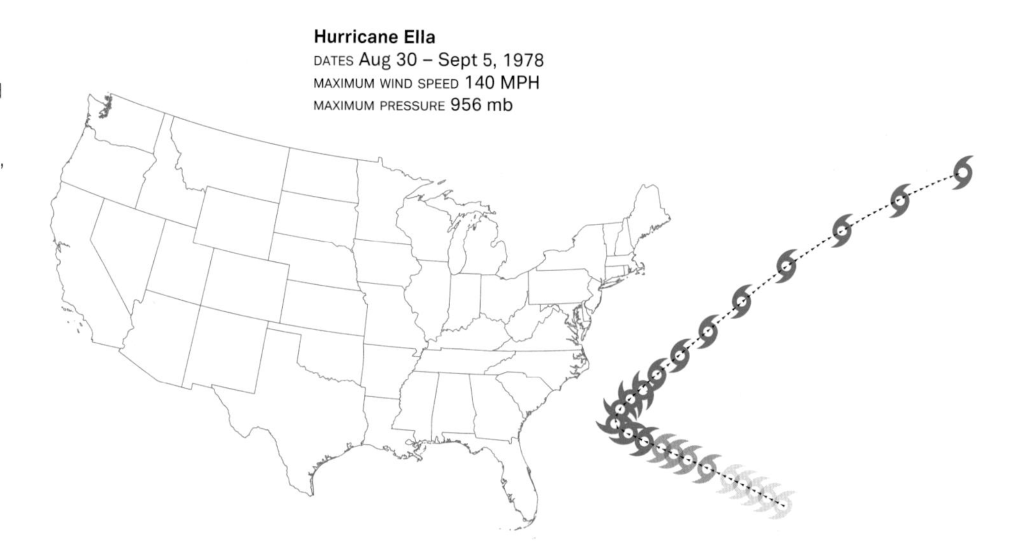

Hurricane Ella
DATES Aug 30 – Sept 5, 1978
MAXIMUM WIND SPEED 140 MPH
MAXIMUM PRESSURE 956 mb

Repairs were undertaken after business hours, seven days a week, welding 2-inch-thick steel plates over the existing connections. With repairs only partially completed, on Labor Day weekend, Hurricane Ella, with peak wind speeds of the order of 140 miles per hour, raced toward New York before, fortuitously, heading out to sea. Needless to say, all of us involved in the repairs stood ready to initiate the evacuation plan. Citicorp, having armed us with beepers, was able to react promptly to alert any and all members of the team. While I can't speak for the others, the passing of the hurricane, well off shore, was a defining moment; an event well deserving of a bit of sleep.

I do recall that thirty to thirty-five welders were involved in the installing of 125 to 150 tons of structural steel, a crew of about the same size as that required for the erection of the building, all overseen by the competent welding inspectors mobilized by our Jim White. The field work was accomplished over a period of about two months, but the most of the work was accomplished during a thirty-day period, attacking the most critical connections first, an achievement accomplished only through the dedication and cooperation of all of those involved in the design, construction, and management activities.

It should be noted that, while both the LeMessurier team and I had uncovered a number of defects not listed in the *New Yorker* article (which article was published nearly two decades after the fact), to the best of my knowledge, all of those defects were properly repaired. Of importance also, in the same *New Yorker* article Bill LeMessurier, perhaps inadvertently, seemed to cite Stanley Goldstein and the LeMessurier New York office as responsible for these significant design errors when, in my view, this assertion was not factually correct. The *New Yorker* article and many other articles in the technical press, hailed LeMessurier for his "honorable" conduct, providing the community of engineers with a hero, something that our profession sorely needed. These articles shed a most desirable and laudable light on our profession but, at least in part, at the undeserved expense of a wonderful engineer, Stanley Goldstein.

On another widely reported meeting with the aforementioned Wriston and Reed, on the lighter side, Mr. Wriston sought a copy of one of my sketches, but all of the copying machines were locked for the night and the administrative staff had departed. Citicorp's chairman was frustrated! My approach was simple: I ripped the door off of one of the locked copying machines and pressed the Power On button. We obtained the required copy, leaving to others the repair of the machine. Perhaps it should be noted that engineers need to learn to act decisively in the solving of unusual problems; for sure, the cost of repairs to the copying machine could not be found in the urgency of the repairs to the building.

I believe that it was mid-September of 1978 when I recommended that the wind engineering effort be concluded and that the evacuation plans be ended.

Were there law-suits? In truth, deliberately, as they might impact Bill LeMessurier or his firm, I learned as little as possible about potential litigation. I recall that Robert Dexter queried me as to the amount of coverage provided under LeMessurier's professional indemnity insurance. While I had that information, I explained that I had no leave to provide LeMessurier's financial details to Citicorp. There have been published reports of litigation and financial settlement; however, I have no knowledge of what actions, if any, were undertaken by Citicorp. Working with our own attorney, I developed a totally comprehensive document, inked by Citicorp, protecting us from all parties, no matter the cause, where even remotely related to Citicorp. In today's world, sans an Earth-moving crisis, I doubt that such an all-inclusive agreement could be obtained from any important corporate organization.

Looking back, it is heartwarming that the two top-level executives, the chairman and the president of Citicorp, maintained such a hands-on posture. They understood fully that this was a completed and operating building, not a parking lot, and that sound engineering judgment should be respected, resulting often in the execution of criteria in excess of that required by the building code. In my view, Mr. Wriston and Mr. Reed, and all of the Citicorp executives, particularly Messrs. DeFord and Dexter, at every step along the way displayed nothing but the highest-level of professionalism and ethical conduct. I remain proud that I had been chosen to stand beside them in this time of strife.

Philip Johnson
1906–2005

Philip Cortelyou Johnson, living to age 98, was an extraordinary architect, a legend in his time. As suited his persona, we were good friends.

"Friends" is surely an exaggeration. The twenty-one year gap in our ages together with differences in our formal education, sexual preference, verbal skills, political and societal persuasions, and personalities may have opened up a kind of dialogue between the two us. The differences allowed me to be playful yet respectful in our conversations. Since Philip had, at least to me, a longer attention span than that exhibited by most architects, we spent many afternoons turning pages of books on architecture, discussing various aspects of the illustrations. I say "discussing" but, of course, Philip dominated our conversations. He would talk, ask questions, sometimes just turn to look at me, and I would have my say. The ending of these "discussions" was almost always abrupt. Perhaps approaching boredom, Philip would proffer a curt "Thanks, Les" and we would part. Still, for me, it was always exciting to learn from this remarkable architect.

One example of playfulness occurred in a discussion I had with Philip and his partner John Burgee. For some reason the topic turned to the problem of headroom in a stair. Philip remarked something not very complimentary about the young architect who had produced the drawings. It became immediately obvious that Philip did not even know the name of that young man, though he must have passed his desk two or three times each day. I could not help but chide him for his thoughtlessness: "Philip," I said, "in not knowing the fellow's name, in not thinking and acting toward him as a talented young architect, you're being a real son-of-a-bitch." John Burgee then chided me for my remark; in turn, Philip, rose promptly to my defense.

Another example occurred over the unceremonious departure of one of the Johnson/Burgee partners, the architect Raj Ahuja. Following his ouster, Raj had come to me asking if we would, temporarily, provide a small office and telephone services. Of course, he and his wife, Savita, being friends and compatriots, I readily agreed. Some weeks later, Monica Svojsik (one of our engineers) and I, talking with John Burgee, were in the Johnson/Burgee offices. As it turned out, this was the precise moment when Philip was in a conference room signing his separation papers from John Burgee. (I was and am an admirer of both men, and knew nothing of the reasons for their separation.) As he finished, Philip, strolling through the tension, walked toward us, with John cutting short our conversation and moving away. Philip and I chatted for a few moments before he said, "I understand that you've taken that son-of-a-bitch (speaking of Raj Ahuja) into your offices." I responded in an instant: "Philip, he's not a son-of-a-bitch, he's a friend of mine." Thinking that I had insulted Philip, Monica blanched. We then spoke briefly on another matter before Philip broke off to don his coat and to depart for a meeting. Monica and I made for the elevator, with Philip immediately behind. Upon reaching the street level, he turned to offer to me a new project. These events tell something of the character of this most complex person, and of my candor in dealing with Philip (and with all architects).

At one point Philip took ill and was confined for about one year to his Glass House estate in New Canaan, Connecticut. Each week I would call, hoping to reach him but Philip's companion, David Whitney, always answered saying, "Philip cannot speak with you at this time." After so many months of such exchanges, David again answered my call, but this time with the message that Philip wished to speak with me. Astonished, I leaped out of my taxi on the way to the offices of Kohn Pedersen Fox Associates, so that I could talk from the more reliable mobile-phone base of a curb-side. After just a very few words, Philip interrupted me to say, "Les, why don't I ever see you?" I thought it better to not mention my fifty-odd thwarted telephone calls. We arranged to meet the following Saturday.

On that beautiful fall morning, it was a more frail Philip who greeted me. He was seated comfortably in the Glass House, but a telltale wheelchair stood in the corner. We spent the morning together, talking of this and that, as though we were old friends. It was wonderful to see his impish smile and to trade barbs with him. On departure, I was exhausted, but feeling warm inside from our time together!

When Philip returned to his offices, starting with one-day per week, he asked me to lunch. Perhaps it need not be said, but, with almost no exceptions, each day Philip dined at the same table in the Grill Room of the Four Seasons, there to be surrounded by his own designs. As several of us entered the restaurant the maître d', stunned to see Philip after his long absence, greeted him warmly, and then went straight to Philip's table, explaining to its occupants that they must vacate immediately even though there were no other tables available in the restaurant. Of course, maître d's are expert in such matters; the exodus took place smoothly and without complaint. As we took our seats, Philip looked around the room in appreciation of all that he saw and all that he felt. Philip was back!

In 2000, the architectural writer Wendy Talarico asked me to consent to an interview with Philip, who was then 94. Knowing full well that he would decline, having refused interviews for some years, I readily agreed. Much to my surprise, Philip acquiesced, telling Wendy, "The only engineer I can stand to work with is Les."

It was in the Glass House, that wonderful creation of Philip's youth, where we met, in Wendy's words, with "the former enfant terrible, the acerbic and wealthy wit, the snappish and irreverent man who cut interviewers into ribbons with his merciless comments."

During the interview, Wendy could not help but raise the issue of the engineer's contribution to architecture. The discussion, as she reported, went something like this:

TALARICO There's a lot of blurring of the boundaries between the engineering and architectural professions. How does that affect architecture?

ROBERTSON I would add art to that, art as a profession. I have misgivings about artists who dabble in architecture. It's okay to be an architect dabbling in art, but not the other way around.

JOHNSON I agree. There are very, very few good architects or artists. I think architects should stop pretending to be engineers too. Calatrava, for example, thinks he's an engineer, but he's more apt to be an artist. And then there's Gehry, a great architect but a true artist. Some people are just born great and he's one of them.

I don't care whether you're an artist, a painter, a sculptor, or an engineer. As long as the key to what you see is an artistic eye. I have one principle in architecture: It must look good. Let the engineer help you with the rest.

R A lot of great architecture has been produced with very mediocre engineering behind it. But you cannot have great engineering without great architecture. An architect can, however, create a wonderful building with almost any engineer. The reverse is definitely not true. For these reasons, it's important that the architect lead in the relationship.

J Oh, but I don't think you can move without an engineer. Who else can tell you the size of a steel beam that won't sag? Engineers aren't involved in projects early enough. That's because architects are too stupid to use them. You have a structure and you want it to be steel. You've used steel before and you think you know all about it. Then you start building and you get into trouble on the trusses and things. Then you come crying to the engineer.

R I don't know about the crying part.

T Screaming maybe, screaming "Help!" I think there's some agreement that engineers are more necessary than ever because buildings are more complicated. And that's partly due to the elaborate designs possible on a computer.

J I don't have a computer. They are inventions of the devil. They are horrible. They don't think. They don't feel. They don't have any knowledge of form. There's nothing like a pencil.

R In engineering, it's important to spend a lot of time in understanding what it is you're building and how you're going to build it, and to delay the computer work for as long as possible. Doing so can mean the difference between an average project and a great project. But that delay tactic is difficult because young people don't know how to do much of anything without the computer.

J They've lost the pencil. Drafting is a lost skill. Dozens of young architects have computer skills, but hardly anyone can draft.

R They've lost the pencil. But when you do get to the point where there's a place for the computer, it's an extremely valuable tool. It allows Philip to change his mind 17 times on the same subject with almost no impact on his or our costs. It allows a kind of refinement that was not so easy to do by hand. And the graphical development of the shape is easier to manipulate and change. You can look at shadows. With a pencil, that is really very tough.

J Except, of course, I don't draw. Never did, never learned how. I have hundreds of drawings, but they're just scribbles. I could get away with that before. But now, especially with new architecture, with all these windows and shapes and materials, everything is a lot more freehand than it used to be, when we had certain rules that we followed, the kinds of rules that governed the design of a cathedral.*

COURTESY OF W. TALARICO

Looking out from the estate to the road, this entry gate has decorative concrete side supports conceived by Philip Johnson. My role was to produce construction drawings and to devise the hoisting/lowering mechanism.

Over the years, as perhaps can be gathered from the preceding exchange, Philip and I had developed our own spirited form of communication. It seems to me, with a comment such as, "Who else can tell you the size of a steel beam that won't sag?" that Philip was just poking fun at me.

Following the interview, Wendy, her photographer, and I walked through a portion of the 47-acre property, slightly melancholic from its obvious need of maintenance. The 1949 Brick House was barely able to receive guests; the 1962 lake pavilion, or Philip's *folie*, had visibly deteriorated; the Painting Gallery, added in 1965, was littered and neglected, and the 1970 Sculpture Gallery cried out for repairs.

As we retraced our steps along the path, passing over a little arched bridge, I stopped to see Philip and David Whitney (they had been together since about 1960) with their backs to us, and with the woods stretched out before them across the valley.

The two men, sitting peacefully together, was a most compelling sight, all seen through the walls of the Glass House. I turned, expecting to see the photographer leveling his camera, but, instead, he was walking briskly away from what was to me an incredible moment. Philip died four years later, in 2005, at age 98; David, 66, followed just five months after.

As an aside, it was David who selected the works of art to be purchased by Philip. In short, David was an art connoisseur of the highest order.

Both Philip and David abhorred uninvited guests to the estate. Philip sketched a sort of gate, designed to discourage intruders. His concept was deceptively simple: a horizontal bar crossing the entrance drive, with decorative concrete work on both ends. I fretted for at least a week, sorting out an economical way to raise and lower that bar.

One evening, the solution burst upon me: the mechanism used in parallel rules! (This was back in pre-CADD days when drafting tables were equipped with a straight-edge bar, or parallel rule, that glided up and down on taut end wires.) I explained to Philip that this was a time-honored solution to the problem. His response was, "Les, why can't I just get out of my car and lift the bar?" "No! No!" I said, "You must operate the bar from within your car, as though it were a garage door." Philip's response was classically "Philip": "Les, I don't need it." My reply was equally straightforward; "Philip, on occasions in your life you must leave such decisions to the likes of me." Philip chuckled, and we parted.

Over the years we accomplished a number of other designs for buildings on the estate, including improved HVAC systems for the Glass House. As I recall, this relatively small project was accomplished by Robert (Bob) Fowler, then a partner in our offices.

More significant was Philip's Studio, a place where he could be completely alone, in a meadow a respectable distance from the Glass House. He sketched a building with just one small room, with a conical dome atop the roof, offset from the center. Philip saw it as concrete, with plaster finishes inside and out. Seeking a more practical and economical solution, I saw it as a mix of concrete block and brick, with a bit of a concrete slab covering the area outside of the dome. Bolstered by the significantly reduced construction cost, all without loss of form, texture, or function, my approach survived, to be published

The Glass House, with Philip leading the discussion at a project meeting inside.

in architectural magazines around the world as an example of the wily Philip, who admired the substitution of the mundane block and brick for the more elegant concrete.

Being constructed over the winter, the contractor fabricated a tent, with wood supporting plastic sheets, allowing the work to go ahead in rain or snow. The neighbors must have been horrified, thinking it to be a permanent sculpture, that they would need to live with this monstrous plastic thing clearly visible from the road. When construction was nearing completion, on a whim, I asked the contractor to remove the plastic and wood in the dark of the night. While I was not present, I was told that he followed my request; in the morning, the neighbors awoke to find Philip's gleaming, white Studio gracing the meadow.

The visual effect of the Studio is striking. From a distance it appears huge with the door sized for a giant. But, planting a foot on the front steps, this "monk's cell," as Philip called it, suddenly shrinks to human scale, an illusion not unlike that of the Pavilion down at the lake which, when viewed from the Glass House higher up the hill, was large, but in reality is not high enough for me to stand erect inside. Also, of passing interest, there are no toilet facilities in the Studio, attesting to Philip's legendary bladder control.

Along the entry drive, from the street to the Glass House, there is a long, curving fieldstone wall, higher than those typically found in rural Connecticut, perhaps five feet high, for privacy. Walking from the Glass House to the Studio, one can see the end of the wall, which end is perfectly square; that is, it is as wide as it is high, making it seem that the entire wall is five feet thick when, in fact, it's much narrower—yet another of Philip's mischievous illusions.

Away from the estate there are Philip's larger projects, some of which are discussed in this book. But before venturing into these big corporate commissions, I'm reminded of a stunning small work: the bell tower at the Crystal Cathedral. The basic configuration, as devised by Philip, remained constant throughout the design: namely, a sort of upside-down crystal chandelier housing a 52-bell carillon, with a chapel in the base below.

John Burgee and persons associated with the Cathedral had developed a marked unhappiness between them, perhaps associated with the design or the construction of the Cathedral. Philip seemed insulated from this tension and I knew nothing of its source. My meetings with Philip, then, seemed to be conducted with the specific intention of isolating John from the process. On a few occasions, Philip and I met in the public hallway outside of their offices.

Recognizing the earthquake risk associated with any structure in the Los Angeles area, I set out to devise a design for these mirror-glass plate crystals. Triangular in cross-section, each measured about 30 inches per side, with the anchorage of the glass at the three corners occupying a significant portion of each surface. As well, I was concerned about the wisdom of housing this mountain of glass over the chapel.

Frustrated after several unsuccessful designs, I decided to trash the whole thing and start over. Almost immediately I realized that a polished stainless steel plate, with just three bends, would conservatively solve the earthquake risk while being both simpler and less expensive. But how to "solve" Philip? I had a mockup constructed, including the expansion joints required by the height of the crystals, the tallest soaring 142 feet into the air. To me, it was nothing less than perfect. On showing the mockup to Philip, I was delighted to find him enthusiastic about the use of polished stainless steel as well as about my proposed detailing.

Stainless steel, when bent, loses its polish at the bend points. At night the polished surface of each crystal facet visually recedes, while the three corner-bends pick up light as bright shafts running the full height of the crystals. It was wonderful! I must confess that I did not anticipate this nuanced effect, a delightful gift. Philip, the wiser, may (or may not) have anticipated this wonder.

While in Europe on other matters, I drove to the Royal Eijsbouts Foundry in Holland, which was producing the carillon for the cathedral. Of interest, though they had fabricated bells for sites around the world, their staff had little knowledge of seismic design. Retiring to a little space they provided, I drew the needed seismic-resistant mount for each of the bells. By accident, I discovered that the foundry was also producing the bells for I. M. Pei's Tower of Angels carillon in Japan so, while waiting for the production of our shop drawings, I sketched new seismic mounts for I. M.'s bells as well.

Coincidentally the staff engineer who, amongst other projects, accomplished for us the most of the analyses for our bell tower, came to us from Japan as an intern. Such interns, ineligible for a work permit and staying with us for just a year or two, are paid directly by their company (as required by US law); it follows that only the best and the brightest are sent. We have enjoyed many such interns but Hideo Nakashima, from Shimizu Corporation (Shimizu Kenetsu Kabushiki-gaisha), was exceptional. With his internship completed just as construction of the carillon was finished, he returned to Japan through Los Angeles, there to view the finished work. Once back home, he wrote to say that he had not fully understood the design until he was able to absorb the completed structure. Of course, realizing that Nakashima-san had learned much while with us, his letter brought both joy and satisfaction.

Philip Johnson was so much smarter, more verbal, and more captivating than this structural engineer. I worked on more than a dozen projects with him, sometimes, as well, with his successive talented partners John Burgee, Stephen (Steve) Achilles, Raj Ahuja, Jeff Sydness, and Alan Ritchie. Alan remained as a partner with Philip, and the name of the firm changed to Philip Johnson Alan Ritchie Architects, which continued after Philip's death.

I so wish I had met Philip earlier, so that I might have participated on even more of his projects and spent even more time chatting with him. It was a privilege and an honor to have touched his life.

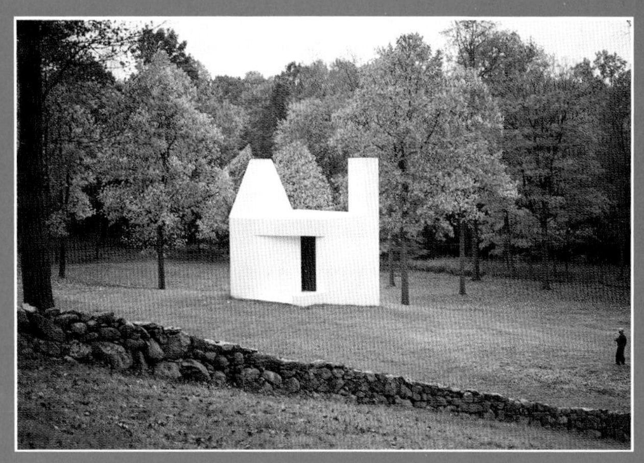

The "square" end to the lovingly detailed stone wall, with the Studio behind.

AT&T Headquarters

NEW YORK, NEW YORK. 1977

Philip Johnson, with his legendary verbal skills, seemed to have firmly positioned himself as the leader, and certainly the most outspoken proponent, of American architecture of his generation. While the architect-theorist Adolf Loos maintained in his *Ornament and Crime* manifesto of 1910 that architecture had outgrown ornament, and Philip certainly was familiar with this influential polemic (he had after all, insisted that "You cannot not know history."). Philip used the AT&T Building to embrace a new and contemporary form of ornamentation. In the process he upended the profession.

Replications of Philip's drawings (actually made by staff as he rarely drew anything) appeared on the front pages of major newspapers and in magazines around the world. It was the broken pediment at the top, which the press derisively labeled "Chippendale architecture," coupled with the architect's own renown, that caught the attention of the press. Later, Philip called the project his *capolavoro*, the culmination of his life's work.

For me, the project started when Philip's partner, John Burgee, armed with a typed sheet of questions and room for comments, came to our offices in order to evaluate our capabilities. First, he informed me that we needed a black/minority associate engineer, who had been selected to be Leroy Callender. I can only surmise that Burgee was skeptical of our capabilities but was propelled by a push from Philip with whom I had met on a variety of architecturally oriented meetings and the like. Of course, even to be considered for the work, I was both excited and pleased.

Following some tough financial negotiations, interviews with folks from AT&T, discussions with Leroy Callender and with the associate architect Harry Simmons Jr. (who perished in 1994 in an airplane accident) and others, we were awarded the commission. A sticking point was my insistence that we be given complete control over all aspects of the structural design, which was resisted by the Leroy Callender

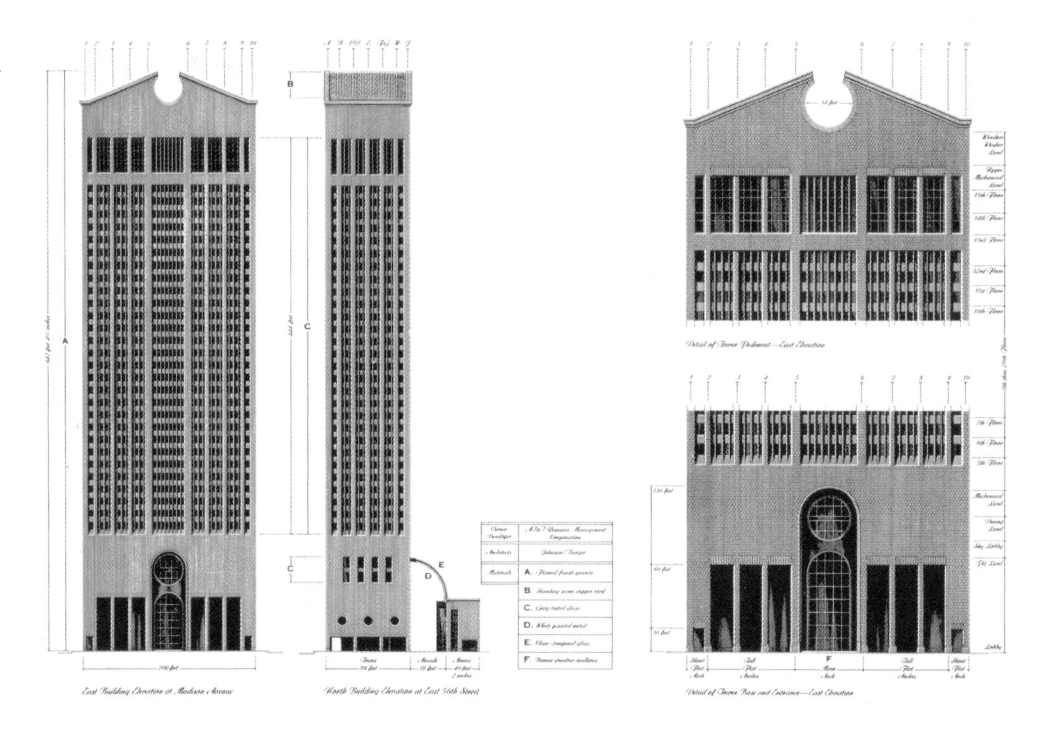

Slightly edited drawings of the AT&T Headquarters; the originals were publicized worldwide in advance of the building's completion.

firm who, understandably, sought a larger role. Ultimately, with the acceptance of Leroy, this matter of jurisdiction proved to be easier to overcome than I had anticipated.

In the original design, before Sony took over the building post-construction and refitted it with ground-level retail, the facade rose from Madison Avenue in a great open arcade. Philip was concerned that the 7-story columns would be somehow unsafe. He said, as he typically did when he pushed boundaries, "I know that I'm being 'bad' in making the columns so tall." Surprising to me, it took several meetings with Philip before he became really comfortable with their height. Indeed, since I assured him that, structurally, the columns could be even taller, Philip grew them more, ultimately raising them to 98 feet.

Philip delighted in the fact that the IBM Building, just to the north, across 56th Street from our project, had been set back from the Madison Avenue property line. Philip seized this opportunity to make his own building more dominant by pushing it right to the edge of the sidewalk. This circumstance opened up a kind of dialogue between the two of us, with my seeking to set the building back enough to install the basement waterproofing and to clear the sidewalks during construction while he remained firmly convinced that we would overcome such difficulties. His persistence prevailed as we were able to convince the contractor that he could apply the waterproofing materials to the surrounding rock, and to then cast the concrete against it, a tactic that I had used successfully on other projects outside of New York City.

It was Philip's desire to have very small joints between the large blocks of granite cladding so that the building would appear more monolithic, thereby dramatizing its radical departure from the glass-and-steel towers that were then the norm. To accomplish this, I devised a support system which, under the lateral motion of the tower, allowed the granite slabs to tilt, rather than move laterally with respect to adjacent slabs. Unfortunately, this resulted in our having to detail the supports for all 13,000 tons (nearly 160,000 cubic feet) of granite, resulting in our expending countless hours in the detail design of the supports for the granite. There were so many conditions that needed careful design, and without additional fees! Additionally, I designed the system so that the failure of the support for a few granite blocks did not result in the catastrophic progressive failure of the granite below. Still, we were able to achieve the very small joint size that Philip enjoyed pointing out to admirers, critics, and others—but at the expense of a rather large dent in our firm's dollar profit on the project.

While the ground-floor columns were very tall, we had no difficulty in making them rather smaller than was shown in the architectural drawings. When I pointed this out to Philip, he responded that he rather liked the stalwartness expressed by the larger columns. It followed that there was a larger-than-normal gap between the struc-

Philip wanted to minimize the size of the joints in the facade granite but realized that, stimulated by wind or earthquake, a floor moves laterally with respect to the floors above and below. These exaggerated drawings illustrate how lateral displacement of the stone joints was reduced, making the joints as small as practical while allowing for the thinner caulking.

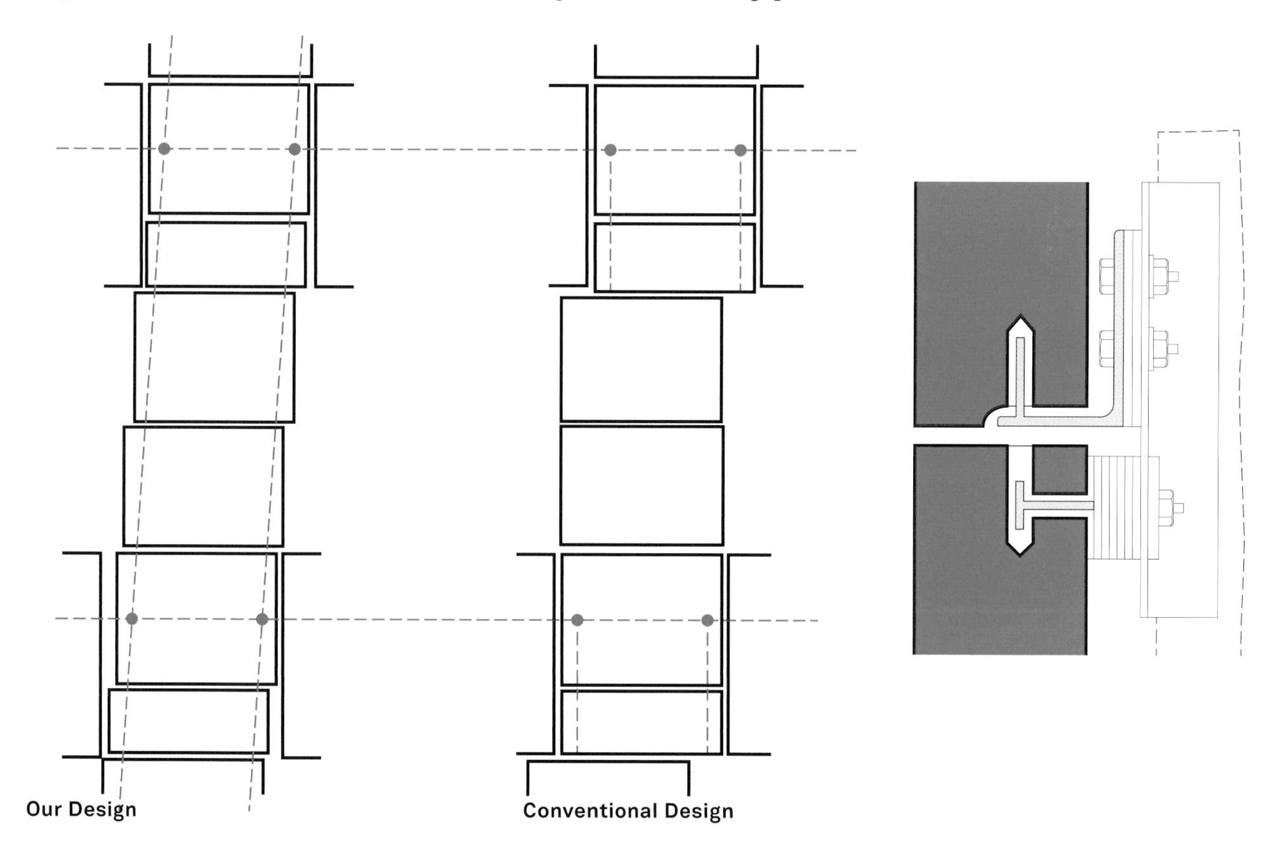

The broken pediment that forms the roof begins to take shape at this stage of construction.

ture and the granite work, which gap ended up carrying piping and the like between the basements and the mechanical room located above the lobby.

On opening night, walking through that mechanical room with the lead architectural representative, I noticed that there was an unsealed gap between the piping from below and the floor of the mechanical room. As the floors of the mechanical room were washed, it was inevitable that water would pass through the gap to the space between the granite and the column, collecting on the ground floor. It was a small thing but, for whatever reason, it had not crossed the minds of Philip's fieldstaff that, to preclude winter freezing of the water, it was important that this area be drained. We went immediately to the ground floor where the lead architect cut away a portion of the caulking at the base of the granite and, sure enough, water poured out. He shook my hand before setting out to provide a permanent repair. This is just one example of my repeated experiences in having architects and contractors fail to provide drainage for unanticipated water.

The broken pediment at the top of the building, which had created such turmoil in the world of architecture, was, for us, a straightforward matter. A greater challenge was the copper roof, with its complex geometries and high wind-induced uplift pressures (suction). On more than one occasion, this created the furrowing of my brow as I sought imaginative support systems for the thin copper roofing. I came to understand that, to produce an economical design, we needed to prevent air from entering under the copper roof as the wind-induced suction attempted to raise the copper panels.

The overall structural system that we employed for the building is very New York. That is, we used every nook and cranny of the 100 feet deep, 200 feet long, 647 feet high building to create the needed resistance to wind and earthquake. Included in these nooks and crannies was the use of concrete-clad steel-plate shear walls, notable because the system was rather thinner than the steel bracing or concrete walls of the past. As well, the system provided sound support for the granite cladding of the lobby, reduced construction time, and saved money. The concrete overlay worked perfectly for the support of the granite but also added stiffness to the steel-plate shear walls while eliminating the need for welded stiffeners.

Much to my delight, the noted architectural critic, Paul Goldberger, used a photograph of our structural steel forming the pediment on the cover of his book *On the Rise: Architecture and Design in a Postmodern Age*. The cantankerous Philip, jokingly of course, chided me for stealing Goldberger away from his "architecture."

Taken from the rooftop of another AT&T building at 195 Broadway, a bronze sculpture, by name Golden Boy, was to be placed on a tall pedestal in the lobby of our new building. The sculpture, I was later told, weighing some 32,000 pounds, was

Bottom Sketch of the openings proposed to lighten the webs of the arcing girders over the through-block arcade.

Opposite A portion of the great structural steel arch over the entrance to AT&T Headquarters.

second in size only to the Statue of Liberty. From our analyses of the wind loads that had been resisted by Golden Boy, I knew that the lower portions, at least, were solid bronze; the museum specialists thought it to be hollow like the Statue of Liberty. Upon investigation, the lower portions of the sculpture, the legs, proved to be solid bronze just as we had surmised, thus authenticating the support system that we had designed.

The through-block arcade and the Annex Building, now greatly changed with new ownership, provided the opportunity to work with architects at all levels within the firm. But the never-to-be-forgotten memory for me comes from a brief discussion with Philip regarding the design of the steelwork supporting the glass roof. Philip saw the steel as simple I-shapes arching above the space below. I saw it, and so proposed to Philip, as having the solidness of the beam webs softened with curvaceous openings. The ever-impish Phillip asked about the shape of the openings. I responded with a sketch of a simple-but-complex shape that I may have seen depicted somewhere. Phillip terminated our discussion by saying "That's fine. Just draw it up." We parted. Both of us, I believe, were well satisfied with this (and many other) experiences.

Over the course of the design and construction of the AT&T Headquarters building we made many friends, from within AT&T, with the contractor and subcontractors, the many consultants, and others. The project was not managed in the usual corporate style, but rather with a more casual approach to management. It was, I believe, Alan Ritchie who made it all come together. Alan, one of Philip's and John Burgee's partners, was ever-present in project meetings, casual sessions, and the like. Alan, his wife, Rosa Hurtado, and SawTeen and I became good and lasting friends.

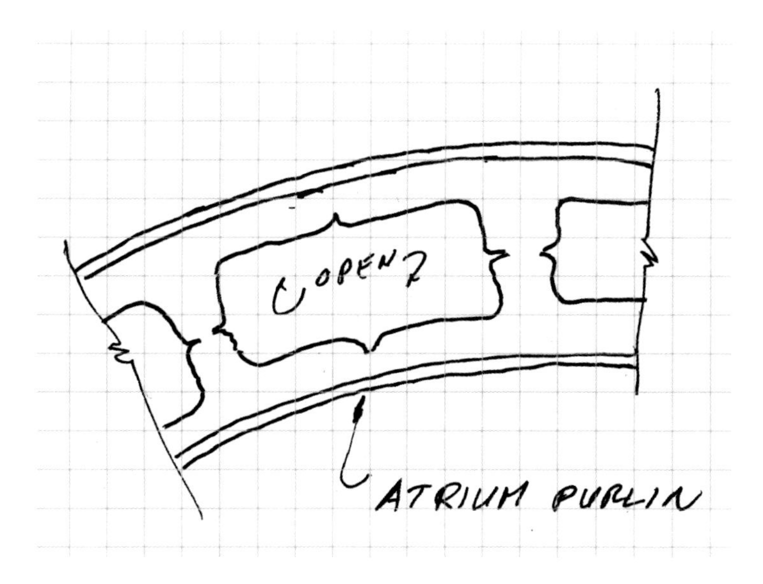

NORTH

One PPG Place

PITTSBURGH, PENNSYLVANIA. 1978

Fascination with PPG's 40-story tower, a landmark project designed by Philip Johnson and John Burgee in Pittsburgh's central downtown, is rooted in its architect, in the undulating glazed facade, and in its 24 spires (with an additional 267 spires located atop the five lower buildings). I'm told that the project has almost one million square feet of glass. While the all-steel structural system was more-or-less straightforward, even so, excitement was to take place!

Following the completion of the structural designs and the letting of the structural contracts, John Burgee asked if we would review the design of the typical aluminum mullion. While outside our contractual scope of work I, of course, readily agreed, asking that the architectural and the shop drawings be sent to us. Upon review, it became clear that the client, PPG, was itself the detail designer of the curtain-wall system and had subcontracted the design to a consulting firm in Texas. While, in my experience, this is unusual, there was nothing in the arrangement to cause concern.

Because of the complexity of the folded facade, with its many interior and exterior corners, I sensed why John had requested our input. I asked one of our staff engineers to review the design; in a half day or so he reported that it was sufficiently conservative. Even so, on a hunch, I asked one of our Associate Engineers for a second review, without providing the first engineer's calculations. Once again, the design of the typical mullion won acceptance. I sent the drawings back to Johnson/Burgee's office, so stating.

Still, while pressed on other matters, something in the back of my mind continued to be concerned. I asked SawTeen See, the most talented young engineer in our office, to check the drawings for a third time. In due course, and very convincingly, she explained that the asymmetrical properties of the mullion had been incorrectly calculated, that the mullions were not capable of supporting the imposed loads.

Since the aluminum for the mullions had already been extruded, this was a very serious matter. I hid away from all outside interferences and reviewed the design myself to find that SawTeen was, indeed, correct. I carried the bad news to the architects. Following detailed discussions with John Burgee, with some participation by Philip, we decided to reach PPG by telephone, with John speaking from our end; I asked that he emphasize the need for PPG to accomplish its own review of the design. The ensuing communication was short and tense. While not being an attorney, it seemed to me that the responsibility for the deficiency and all of its consequences rested firmly with PPG, not with the design team.

After some review by PPG's consulting firm in Texas, the mullions were melted down, the metal reused, and properly sized mullions newly extruded. Thereafter, construction of the exterior wall proceeded uneventfully. We did not approach the curtain-wall engineers, and were not asked to recheck the revised design. Of interest, some 25 years later, in a design meeting on another project, an engineer from that firm approached SawTeen, identifying himself as the engineer who had made the error in the original calculations.

But wait! Yet another adventure was to take place. At the level of the entrance lobby, the exterior wall was set back from the building above providing, at plaza level, an out-of-doors overhang on all sides of the building. Under the requirements of the Pittsburgh Building Code, this outdoor area must be protected by sprinklers, a straight-forward design problem handled by the plumbing engineer, certainly not by the structural engineer. Rather than heat the ceiling space to prevent water from freezing in the sprinkler piping, a dry system was selected; water flowed only when activated by the heat or the smoke detectors.

A post-construction test confirmed that the sprinklers worked properly: the water pumps had been turned on, filling the piping and spraying the space below. The system was then turned off. To prevent freezing of the water, it was essential that the piping be drained. A technician opened the valve so that the water could be wasted into a maintenance sink, located in a mechanical room. The water came out, continuing to drain well after the technician had departed at the end of the workday. When the pipes were empty, the flow of water stopped. All was normal.

But during the night, perhaps it was as little as a spider getting into one of the smoke detectors, the alarm sounded, and the fire pumps sprang into operation, with the drain valve still wide open. Water gushed out of the mechanical room, pouring down the stairs to the basement. The transformer room flooded, causing the cessation of electric power to the building. There was no reason for me or anyone in our firm to be involved, and I was not informed of the incident until receipt of notice from PPG's attorneys that, amongst others, we were being sued for damages associated with the

The complexity of the facade is shown both completed and under construction.

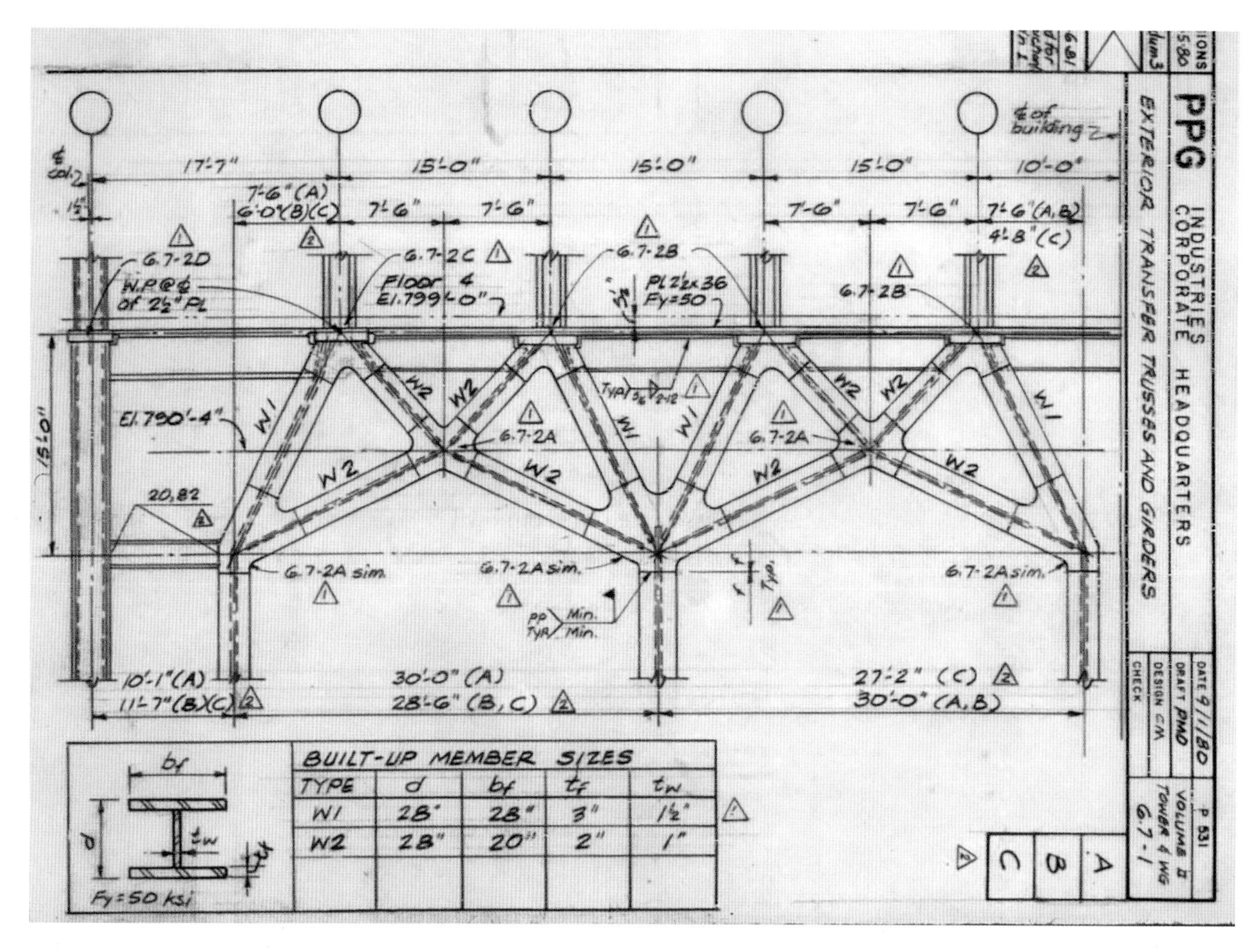

BUILT-UP MEMBER SIZES

TYPE	d	bf	tf	tw
W1	28"	28"	3"	1½"
W2	28"	20"	2"	1"

Above A drawing of the perimeter transfer truss required to follow Philip's design for the entrance area. This design was criticized by the reviewing engineer, who asked, "Why not eliminate the trusswork by allowing the columns to carry through?"

Right A perimeter transfer truss under construction.

event. Pleading with the attorneys proved fruitless. Since the cost of the lawsuit would be significant, I called the president of PPG, reminding him of the good work that we had done on his curtain wall and requesting his intervention on our behalf. He promised to look into the matter.

Almost immediately, I received a telephone call from PPG's lawyers: "How dare you communicate with PPG without our permission!" By then, being moderately angry over such high-handedness, my response was quick and short: "It was easy." I hung up the telephone, and cut off the conversation without waiting for a reply. Within a week, the frivolous law suit against us was dropped, and I called PPG's lawyer to apologize.

Yes, there were other less exciting adventures but this provides some insights into our involvement into the many issues not associated with our own contractual obligations. As well, they demonstrate something of the personal participation of members of our firm in the desire to create wonderful buildings, not just wonderful structures.

Puerta de Europa

MADRID, SPAIN. 1988

These wonderful buildings, leaning toward each other over a roundabout of Paseo de la Castellana, one of the most important avenues in Madrid, are believed to be the first modern multi-story buildings designed deliberately to cantilever from their foundations. Philip Johnson and John Burgee, with Pedro Sentieri, received the design commission. We were under contract to Johnson Burgee Architects.

Earlier, in the mid-1980s, working with the architect Gunnar Birkerts, we had nearly completed the concept designs for an unbuilt leaning building in Michigan: Domino's Farms Tower known by us as Domino's Leaning Tower of Pizza. A cathedral was to crown this 30-story high-rise. Also, several years before, working with the architects Minoru Yamasaki, Jorge Mir Valls, and Rafael Coll Pujol, we had designed the then tallest building in Spain, the Torre Picasso, 157 meters in height and also on Paseo de la Castellana.

Torre Picasso and Puerta de Europa were sponsored by related development groups with whom, more than a decade later, in association with César Pelli and Diana Balmori, we designed the structural systems for a high-rise in Argentina: Repsol-YPF Tower (2008) in Buenos Aires.

How did the concept of the leaning buildings arise? There being two major developers, each requiring office space and a strong individual identity, the construction of two towers was virtually a given. Because of the presence of below-grade rapid transit, it was virtually impossible to construct the towers close to the Paseo; a setback from the street was inevitable. I'm told that, at the initial planning session, a two building scheme separated by the roundabout did not generate the desired sense of uniqueness and excitement. Phillip placed his elbows on the table, with his arms vertical. Then, silently, without words, he moved his hands toward each other, demonstrating the concept of two leaning buildings — just one of Phillip's brilliant ideas. Well, Phillip is quoted to have later said: "We must end the right angle if we do not want to die of boredom. The skyscraper is over, we can forget it." Without question, for Puerta de Europa, there was no looking back.

Philip's idea was quickly developed into conceptual exterior elevations, with each building sloping 15 degrees toward the other, forming the *puerta* or "gate" to Europe. The leading edge of each building was to be 45 millimeters, roughly 2 inches, inside of the property line. Clearly, the cantilevered towers could not extend beyond the boundaries of the site; 45 millimeters, then, was the largest permissible lateral deflection.

It was at this point that we were first introduced to the project. The architects envisioned an architectural expression on the facade with single verticals on the two sides of the building and a single diagonal on each of the two overhanging facades. The single vertical on each of the two sides offered little to us, but the diagonal expression on the overhanging faces opened up an opportunity to improve the reliability of the structural system. However, under gravity loads, the use of single diagonals would cause a permanent gravity-induced lateral deflection and twist of the building. Of course, all of this is possible but a change to crossed-diagonals would eliminate this lateral thrust while increasing the reliability of the determination of the lateral deflections and significantly increasing the torsional stiffness and strength of the towers. While Philip quickly accepted my proposal for crossed-diagonals, John was a tougher sell, though he did acquiesce.

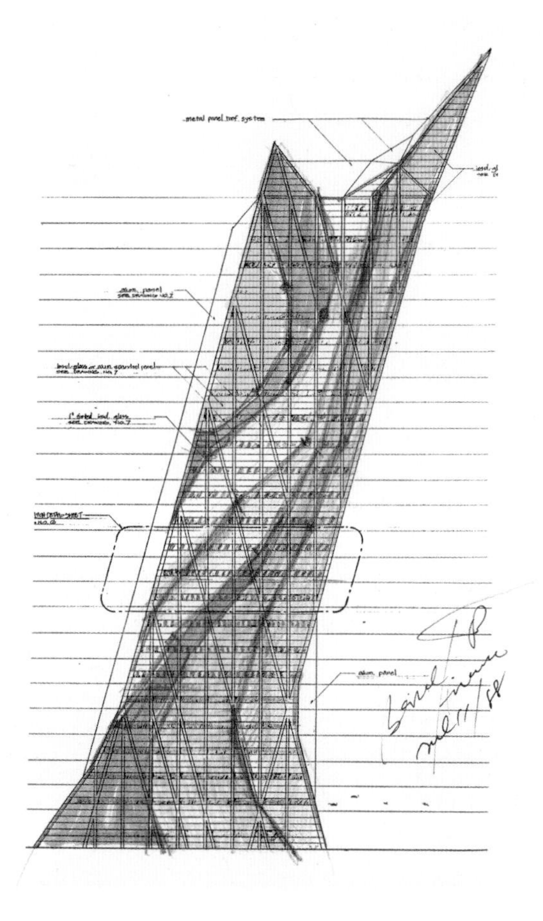

Top Gunnar Birkert's study for a leaning building, for which we provided the structural engineering.

Above The finished 26-story towers of Puerta de Europa.

While these verticals and diagonals were replicated by us into the structure, other vertical elements of the structure of the perimeter facade are conspicuously not expressed in the architectural facade. There are those who consider this to be "dishonest" but, for me, there is the full realization that these "gates" to Europe need express elegance and simplicity, making superfluous any such structural dishonesty. Keeping the weight of the cantilevered portions of the buildings as light as practical provided a solid philosophic base for the selection of construction materials. We used slip-formed concrete cores, which were to remain vertical inside of each sloping building; for the office floor framing we used steel beams carrying a profiled metal deck overlain by a concrete slab. A heliport covered the mechanical equipment at the top of each building, painted blue on the west and red on the east for easy identification by pilots. We used the spaces below the heliports, occupied by the mechanical and electrical equipment, to house deep truss-work.

Of interest was the fact that Philip, once having established the basic architectural concept of two leaning buildings, did not seem to participate in the development of the project. Nonetheless in Spain, and indeed elsewhere around the world, Philip is routinely thought of as the architect; having proposed the concept, it is a title that he richly deserves. While accomplishing the vital role of staying on top of the design, John Burgee was not overtly active in the detail of the design, although his hand is certainly visible. It was John who insisted on the use of thin stainless steel wall panels on the facade, thereby expressing the diagonals and verticals, an unfortunate decision as the too-thin, out-of-flat, stainless steel wall panels detract a bit from the overall appearance of the buildings.

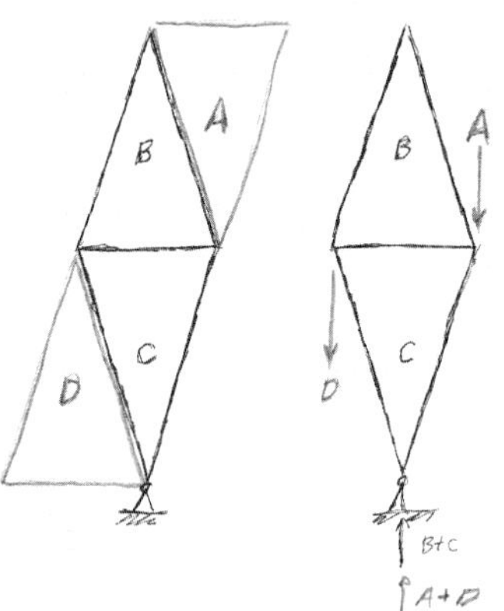

Some believed it was not possible to design a leaning building that does not deflect laterally under its own weight. This sketch, showing the balancing of gravity loads, proves otherwise.

On the whole it was Steve Achilles who carried the brunt of the architectural work, travelling regularly to Madrid, attending meetings and the like. Steve was the Project Manager but, for a variety of very good reasons, his role expanded to include essential portions of the design. Working in three languages, Steve was able to efficiently manage the work in the office while coordinating with Pedro Sentieri Cardillo, the architect of record, with the other Spanish consultants, and the client, Kuwaiti Investment Office, getting the project built.

Our first task was to develop a structural concept that dealt with the deflections associated with the lateral cantilever as well as the burden imposed by the 45-millimeter limitation. I developed several systems that allowed the cantilever but without having gravity-induced lateral deflections created by that cantilever. Only one solution is shown here. All of these systems suffered from a lack of elegance, being clumsy in appearance, making them unacceptable both to me and to the architects.

The towers under construction on the Paseo de la Castellana, forming a gateway into Europe at the northern edge of Madrid's business district; the tall cranes, not being laterally attached to the buildings, were cantilevered from their foundations.

Of course, the structure could be cambered, that is, constructed out-of-plumb in a direction opposite to the gravity-induced lean of the buildings. The other approach would be to provide a force opposite to the direction of the cantilever, balancing the natural tendency of the structure to deflect.

Because, in Spain, concrete was the material normally used for structural systems, and because the use of concrete for the services core was philosophically sound, I was reluctant to develop an all-steel structural system. However, we were able to prove to ourselves that, with cambering, it was not practical to reliably generate concrete with the necessary long-term physical properties to attain the 45 millimeter limitation.

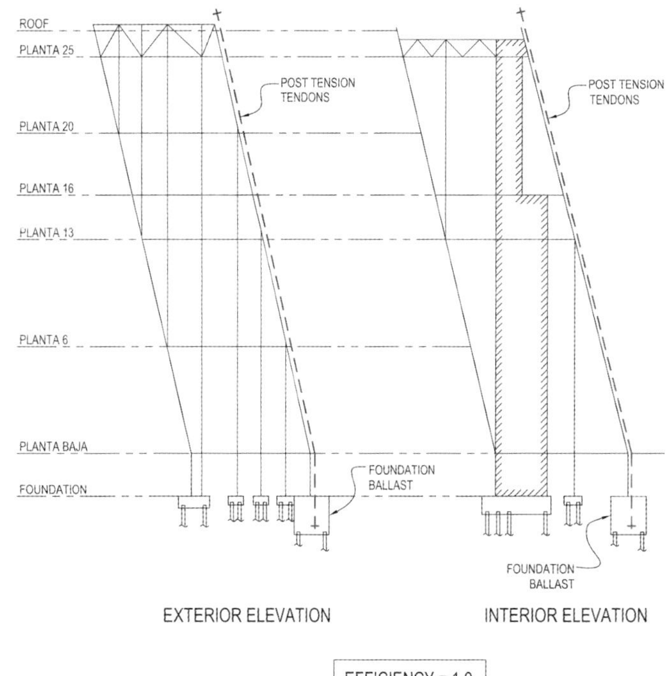

Eccentric post-tensioning, the solution to the lateral displacement of the leaning high-rises buildings.

The better solution, that of providing a lateral force able to counterbalance the effects of the cantilever, was the simplest solution to an otherwise complex problem. Albert Einstein is credited with saying "Everything should be made as simple as possible, but not simpler." In this case, the simplest solution was to allow the buildings, during construction, to temporarily deflect past the 45 millimeter limit, and to attach post-tensioning cables to the top of the building in order to pull the gravity-induced lean of the building back to the vertical.

My approach was to provide post-tension cables in line with the concrete core and with the structural steel in the two planes of the building's side walls. But wait! These cables imposed an additional vertical load onto the structure. Subjected to these additional loads plus the gravity-induced vertical loads, the concrete cores must shorten over time, forcing the buildings to deflect in the direction of the cantilever. However, by deliberately attaching the post-tension cables eccentric to the axis of the concrete cores, the resulting moment, with time, will move the building away from the direction of the cantilever. By judiciously adjusting the eccentricity of the connection of the post-tension cables to the services core, it became possible to set these two deflections equal to each other, resulting in no tendency of the buildings to move laterally over time. The variation in the strength of the concrete, being associated both with bending and with axial loads, had little influence on the determined eccentricity.

All of the above, while theoretical in nature, proved in the field to be both practical and simple. The towers, then, were deliberately constructed without cambering; only when the structure of the building had been completed, were they pulled back true and plumb by the post-tension cables. Laser plumb-bobs were used to authenticate our theoretical predictions.

The post-tensioning system was taken from the standard vocabulary of the concrete industry. Physical protection for the tendons was provided by threading them through heavy steel pipes, followed by grout-filling those pipes.

During the post-tensioning operations, one of the wires broke. Impelled by its internal strain energy, the tendon curved overhead, before passing completely through the body of one of the workers. To our amazement, the wire was cut, freeing the workman, who walked off of the site, both ends of the steel wire projecting from his body, to be driven to the hospital. As I understand it, in a few weeks he was back on the job! Tough, these Spanish construction workers.

The project, of course, needed to pass the scrutiny of the Building Department. The mayor of Madrid, taking a personal interest in the design, asked to meet with me. While he spoke little English and I no Spanish, we were kindred spirits, able to communicate. Understanding the use of post-tensioning, perceptively, he asked how the wires were to be anchored into the ground. I explained that two systems were under study: the use of ground anchors which would require the cables to be carried deep and anchored into the underlying soils or, the construction of large, 15,400 ton, cast-in-situ concrete blocks as ballast for the anchoring of the cables. The mayor, with essentially no hesitation, said "You shall use the concrete blocks." This being my preference, I immediately agreed and we received our Building Permit.

The concrete block solution mandated coordination with the underground transit system. Engineers for these public transportation systems are appropriately wary and conservative. Even so, perhaps associated with the mayor's approval, these obstacles proved to be easily hurdled.

For buildings of moderate height, it is customary, and proper, for the structural engineer to accomplish the structural analyses as though the structure was to be completely constructed on the moon, i.e., without the presence of gravity, with the application of gravity delayed until the completion of the computer coding. For a variety of reasons, this could not be done for these leaning buildings. Instead, we specified a rigorous construction sequence:

- Construct the concrete walls of the services core so as to stay well ahead of the erection of the structural steel;
- Erect the structural steel and profiled metal deck for ten floors;
- Concrete the top-most floor of this ten-floor package;

- Proceed upward with the structural steel for another ten floors while concreting the unfinished nine floors in the package below.

The above sequence was repeated to the top of the building. Careful surveying, following our specification, was accomplished to ascertain that the ensuing lateral deflections were consistent with the theoretical predictions.

Monica Svojsik of our office agreed to be our site engineer during the construction phase. Turner International had been retained previously to provide construction management services but the man sent to the site, while competent in his field, did not speak Spanish; both he and Turner were quickly terminated from the project. Monica, by contrast was fluent in Spanish, having lived much of her life in Lima. She was competent and personable, tall, and beautiful and, best of all, able to command the respect of both the superintendents and the construction workers. She proved to be the perfect site engineer.

There was yet to be another memorable moment in the history of this project! Phillip Johnson and John Burgee, for reasons of their own, had dissolved their partnership. John, then, held our contract. While Puerta de Europa had been completed, he was significantly behind in his payments to us (a scenario all too common among architects, who tend to be terrible bookkeepers). John telephoned to say that we needed to talk, offering "I'll come over to your offices." I replied, "It's easy for me to come to see you." "No," he insisted, "I'll come to your offices." In my many years of practice, for the head of an important architectural practice to travel to our offices had almost never before taken place.

We met in the privacy of our conference room. John said "Les, I've been paid, but I cannot pay you." He owed us a quarter of a million dollars which, even in today's economy, was a lot of money to me. Being a bit dazed, I cannot recall the remainder of our conversation in any detail, but I do know that we shook hands, as friends should. The root of all of this is that John had declared bankruptcy. Some days later, other consultants called wanting to band together to seek redress. Being impressed with John's honesty, candor, and frankness, and believing that he had spoken the truth to me, and having no desire to sue, I declined to pursue the matter with them and have never regretted my decision. Indeed, later, John applied for a position in Washington, DC as Architect of the Capitol, receiving my enthusiastic letter of support.

It should be said that, as each of the several younger partners left the firm, we have remained friends with them all, including John Burgee.

Kiyonori Kikutake 1928–2011

I met Kiyonori Kikutake obliquely, as he presided over a 1994 jury that had selected César Pelli, Richard Rogers, and me as the first Honorary Members of the Tokyo Society of Architects & Building Engineers. The event was very formal, or mostly so, with the dignitaries in tails and César and me in tuxedos; to my amazement, Richard went casual in sneakers and red socks. We arrived on the podium alphabetically, organized by height, and age, César being the tallest and oldest and Richard, the shortest and youngest. The Japanese attendees found this curious coincidence quite amusing.

Despite the formality of the event, the day and the evening were good fun, providing as well the opportunity to learn a little of Kikutake-san, a prominent Japanese architect and one of the founders of the Metabolist Movement. Our first encounter grew into a special friendship, with both Kikutake and his wife, Mutsuko Theresa Smith. Of interest, prior to our involvement with the Mori Building Company, Mutsuko-san had provided Japanese/English translations for one of our clients, Minoru Mori; on her marriage to Kikutake, should Kikutake seek a commission with Mori Building Company, there would develop a possible conflict of interest, forcing her to resign.

In 1995, an international competition was announced for the $200 million Saitama Arena in Saitama City, Japan, providing major public spaces to accommodate sporting events, theater, exhibitions, concerts, cultural affairs, and the like. Taken simplistically, Kikutake's interpretation of the competition called for two long-span rooms, one above the other, each approximately 300 feet by 400 feet. The upper room was to contain a garden and trees, a small river and a lake, a theater, several laboratories, and other facilities, all under a column-free roof.

Kikutake asked me to come to Tokyo, to participate in the competition. I was overjoyed. Recognizing that our competitors were some of the most imaginative architects on the planet, this was no time to be timid. My immediate reaction to Kikutake's planning was that it was not sufficiently dynamic. I proposed to align the lower building with the primary city grid as Kikutake had done, but to rotate the upper building in alignment with an adjacent diagonal street grid.

Both Kikutake-san and his designers were enthusiastic, but quick to point out that this would place significant gravity-induced loads from the roof onto the long span of the lower building.

I eased their minds by unfolding my idea for a giant V-shaped roof that would rest comfortably on the edges of the long-span structure below. The 'V' was to be created by two arches, founded at the corners of the building, but being tangent at their crown. The upper roof, then, became a giant, interacting space structure with the loads carried largely by the primary arches to the edge of the lower roof. The transverse trusses act in part as arches and in part as beams, the balance between these two actions being adjusted by the amount of the post-tensioning of the primary arches. Conceptually, the transverse trusses could be brought into tension, creating a beneficial uplift on the long span of the building below. This was a lot for Kikutake and his designers to absorb, but absorb they did, with enthusiasm.

Because of the interaction between the primary arches and the transverse trusses, the level of structural redundancy is high. I believed this redundancy to be essential for an important structure located in a high seismic risk area of Japan; terrorist attack was of lesser concern, but remained a potential consideration. Under this concept, any truss member of the system, including chords of the primary arches, is able to be severed without causing progressive collapse of the entire system, thus providing enhanced fundamental protection to life and to property.

With the goal of increasing the stability of the Teflon and glass roof under the assault of typhoon winds, I proposed the controlled provision of a moderate level of positive internal air pressure in the upper space. Thanks to Kikutake, and only partially understood by me, together we hold Japanese patents on the structural system for the Saitama Arena. While—alas!—we did not win the competition, we took halting steps, built on our friendship, toward a professional relationship. The experience solidified a close bond between Kikutake and Mutsuko with SawTeen and me. I treasure the letters and books we exchanged.

Unfortunately, we did not manage to work together again. While Kikutake-san and I were born in the same year, he passed on December 26, 2011. We continue to enjoy our friendship with Mutsuko.

An overview of the proposed Saitama Arena. It was only the two roofs over the arena that fell within our areas of the competition.

A closer look at the two-roof system. The upper roof is rotated with respect to the lower roof, allowing each to be aligned with the adjacent city grid. The upper primary arches were to be post-tensioned, thus relieving the load on the roof structure below.

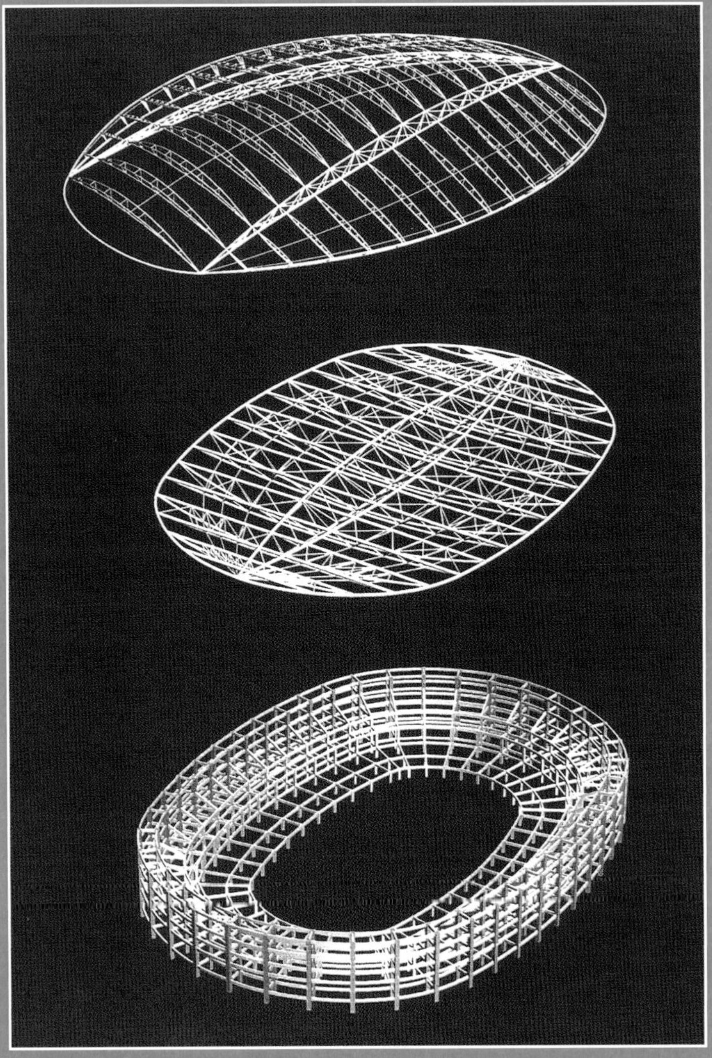

I. M. Pei
1917–

Ieoh Ming Pei, but we always call him I. M., was born in 1917 in Guangzhou (then Canton) but spent his early years in Hong Kong and Shanghai, with the summers of his youth in the legendary gardens of Suzhou. At age seventeen, traveling to the United States, he enrolled in the Department of Architecture at the University of Pennsylvania, moving almost immediately to MIT. On graduation, he enrolled in Harvard's Graduate School of Design, which was then headed by Walter Gropius.

At age 22, in 1939, there was a happy and lasting marriage between I. M. and Eileen Loo. In 1999, SawTeen and I attended a small gathering in honor of their 60th wedding anniversary, with the celebration taking place at their delightful weekend home in Katonah, New York, designed back in 1952 by I. M. We were enthralled by the slide show developed and narrated by members of the Pei family. It was a boisterous and happy event, with each of the non-family attendees being introduced with a short remark. Noting that there was an omission in these introductions, Eileen leaped to the fore so as to personally introduce SawTeen.

Alas! Eileen succumbed to cancer in 2014; the success of her and I. M.'s 75 years of marriage is a yardstick of aspiration for us all. Following Eileen's memorial, where a small group assembled to recognize her life, SawTeen and I wrote (in part):

> *For those of us who were committed to the following of her baton, Eileen was the ultimate director, the ultimate authority. When we would drop in at their house on Sutton Place, it was most often Eileen who would open the door, to usher us into the dining room for a brief visit with I. M. Then, the baton would wave toward the kitchen where veggies were in need of peeling, oysters in need of shucking, or the like. Eileen, of course, was expert in such matters, while this country boy and SawTeen, a Chinese import from Malaysia, could only do our best to perform as commanded. Over the kitchen sink, Eileen was able to carry on a spirited and animated conversation, thus encouraging us to peel that last carrot or shuck that last oyster all under her watchful eye.*
>
> *On another occasion, when he received the Gold Medal from the Royal Institute of British Architects, I. M. suggested that we partake of dinner at a famous Chinese restaurant, a really good idea except that it was the eve of Chinese New Year; all tables were booked, and we had no reservation. On arrival, we leaped out of the taxi, but found ourselves at the end of a long line of people waiting to get in. We stood patiently, perhaps wishing that we were somewhere else. A few moments later, Eileen marched by, giving us a big smile as she wended her way to the head of line. While we have absolutely no understanding of what she*

did or said, we gained instant entrance to a lovely table, with bowing wait-staff.

We are so very proud to have touched the life of this delightful and lovely woman, Eileen Loo Pei.

In 1948, 31 years of age, I. M. joined the firm of Webb & Knapp Real Estate Company in New York City as director of the Architecture Division. He worked closely with William Zeckendorf Sr., who headed the firm, and together they created many exciting projects in cities across the United States.

I. M. became a naturalized US citizen in 1954; a year later, he formed his own firm, I. M. Pei & Associates, but did not leave Zeckendorf officially until 1960. In 1966, he renamed his firm I. M. Pei & Partners, to become Pei Cobb Freed & Partners in 1989.

Upon retiring from the firm in 1990, I. M. continued his practice as I. M. Pei Architect, frequently working with his sons Chien Chung (Didi) Pei and Li Chung (Sandi) Pei at Pei Partnership Architects.

Among the numerous medals and prizes I. M. has received, the Turner Prize for Innovation in Construction is worthy of a few words: The first Henry C. Turner Prize of the National Building Museum and the Turner Construction Company, in Washington, DC was unexpectedly awarded to me. There I met Ms. Susan Henshaw Jones, who had succeeded as president in wrenching the National Building Museum into the contemporary world. Susan and her husband, Judge Richard K. Eaton, became dear friends of SawTeen and me. As names were tabled for the second Turner Prize, I. M. Pei, the originator of many areas of construction technology and design, became a logical candidate; the basic problem being that he was reluctant to accept further prizes. How to convince him that he should make an exception? After forwarding the nature of the prize to I. M., Susan and I arranged for the three of us to dine at Gino's, his then favorite luncheon restaurant in New York, to discuss the matter.

I. M., arriving a bit late, took his seat while greeting us with his wonderful smile. Without waiting for us to present our reasons for his accepting the Turner Prize, quite solemnly, he stated that he had never received a prize for his contributions to innovation in construction technology, and was looking forward to the ceremony!

Having studied at MIT, I. M. enjoyed a remarkable, yet basic, knowledge of structural engineering. Even so, he used this knowledge as a vehicle for evaluating the thoughts that I brought before him. In this way, he was quick to understand our ideas, so as to speed and to improve on project development. Indeed, once brought to his attention, there was no detail too small or too large to escape his perceptive comments. On such occasions when we presented a proposal divergent from that which he had tabled, I. M. would take a moment to consider the alternatives, before agreeing with us or not. His projects as discussed herein offer something of the depth of his understandings.

All of the architects discussed in this book are geniuses in their own right, but I. M. had an incredible talent and good fortune to work with kings, presidents, and developers so as to bring to them an understanding of the nuances and the importance of his designs. I. M. remained supremely confident, or so it seemed to me, that his approach had been extensively researched and was the best for any given project. Still, at times, something would develop that ran hard against his concept, at which time we would retire to a quiet place so as to modify his designs in a way that addressed the newly raised issues.

One such exception was for the conoid, or great glass wall, of Meyerson Symphony Hall in Dallas, wherein he made an instant decision. As shown in the photo at right, at one end, the wall is in the vertical plane and is very short while, at the other end, it slopes and is very long. Concerned with the possible ripple effect under turbulent wind, I pointed out to I. M. that the mullions supporting the long end of the wall would deflect substantially more than would the mullions at the short end, and that the effect could be visually disturbing. Realizing immediately the situation, he asked how to ameliorate these large differences in deflection. I suggested that the introduction of transverse trusses would be the customary solution and that these transverse trusses would readily provide the needed stiffening. In return, he asked where they should be located. I replied that they could go most anywhere and, not wanting to prejudice him, I drew a slanting line across the conoid surface. Without hesitation, instantly, I. M. augmented my line, adding other lines to display lovely geometry.

Morton H. Meyerson Symphony Center. Trusses of differing spans and orientation define the sweeping conoid.

The Museum of Islamic Art in Doha provided other examples of I. M.'s versatility. For example, with the museum to be constructed a short distance offshore, the Qatari authorities decided that a sheet-pile wall, outlining the building perimeter would first be constructed, with the museum construction to follow inside of that wall. I countered with the proposal that, using dredged sand from the ocean floor, a peninsula be constructed encompassing the area of the building plus a bit more than a road-width outside of the museum perimeter. The museum, then, could be constructed following conventional technologies, and with materials and equipment arriving by truck. Upon completion of the museum, again by dredging, the surrounding sand would be returned to its initial resting place, leaving the museum seeming to float in the sea. I. M. leaped at my proposal, giving it his unqualified support.

In the beginning, the Museum's officials insisted that the structural design be completed in strict accordance with British Standards. I explained that we were fully capable of complying with this stipulation but that, were US Standards to be employed, there would be a savings in cost and in time of construction. While I stressed that the US Standards were preferred, the officials remained firm, insisting that the British Standards were to govern. Later, in a larger meeting, with I. M. present, the Museum officials raised the matter again. In an uncharacteristic manner, I. M. rose, requesting justification for the Museum's pronouncement. There followed a short recess after which the meeting resumed with the opening announcement that the US Standards would be mandated.

A third example of I. M.'s approach is provided by the Rock and Roll Hall of Fame and Museum. With the auditorium carried on a single concrete column, disappearing into the water, I was concerned that oil slicks, so common in Lake Erie, would stain and disfigure the column. I awoke one night with the solution! I went to Nippon Steel, a large Japanese steel com-

Above The elegant Museum of Islamic Art, Doha, United Arab Emirates.

Right and opposite Inside the glass tent of the Rock & Roll Hall of Fame and Museum, and I. M. Pei outside the building on the shore of North Coast Harbor. When these two projects are seen in juxtaposition, something of the breadth of Pei's architecture is realized.

pany with the proposal that they provide a titanium sleeve inside of which would be cast the concrete column; in return, they would be provided suitable recognition in the program for the opening ceremonies. It would be wonderful to have the world's only titanium column! My proposal was accepted by the Japanese steel company but Alas! for unexplainable reasons, our contractor refused this offer. I took my idea to I. M. who, not wishing to invade contractor territory, chose to not get involved. But that we could have had that titanium column!

I. M.'s design called for a sloped glass skylight at the front of the building to showcase and illuminate the multi-level lobby and circulation spaces inside. The large triangular skylight required long-span trusses for support. I organized those trusses in a triangular pattern so as to reflect the grandeur of the space while providing bracing in the plane of the skylight. While I planned to place each truss perpendicular to the plane of the glass, I. M.'s staff insisted that they be in the vertical plane. My desire was based on a reduction in cost and depth of truss and because, from inside of the great space, trusses in the vertical plane would appear to be larger than those perpendicular to the plane of the glass. Not wanting to lose this battle, I went to I. M., explaining my reasons. He understood perfectly my logic, with the end result being that the trusses were changed to be perpendicular to the surface of the glass.

The point, here, is that I. M. was quick to make good use of his engineering background to improve on his designs and to make them more affordable.

Over the years, we worked with I. M. on many projects, only some of which are discussed on the following pages. Perhaps of interest, there were times when we simultaneously had contracts with I. M. Pei & Partners, Pei Cobb Freed & Partners, I. M. Pei Architect, and Pei Partnership Architects. This long relationship, and the many projects born of it, continues to be appreciated and honored by me and by all of our staff who have worked with I. M. Pei.

Kapsad Development

TEHRAN, IRAN. 1975

I have no knowledge of why I. M. contacted me; perhaps it was a recommendation from the well-known architect Jaquelin T. (Jaque) Robertson, who had experience in Tehran, or from Bartholomew T. (Bart) Voorsanger of I. M.'s firm. In any event, in 1975, I. M. telephoned, asking that I stop by to discuss a high-rise project, perhaps 300 or 350 meters high, to be constructed in Tehran. Of course, I did my homework prior to meeting with him. The following is a brief summary of that which my efforts uncovered:

Lying at the base of the Alborz Mountain, Tehran is known to fall in an area of high seismicity. Sahr Rey, a southerly suburb of Tehran and once the capital of Iran, had been devastated by earthquakes in the past. Tehran, Iran's capital city since 1795, had not experienced a devastating earthquake since the beginning of the nineteenth century. The earthquake faults in and around Tehran had been mapped by British engineers and geologists.

The Sahneh Earthquake of 1957 and the Buyin-Zara Earthquake of 1962, both shallow of magnitude 7, produced ground tremors in Tehran of intensity 5. Since current construction in Tehran was shockingly devoid of seismic resistance, even those moderate earthquakes, if either had been centered significantly closer to Tehran, would likely have produced catastrophic destruction in the city. Earthquakes of magnitude 7.5 or larger were believed to be expected from the North Tehran Fault located within 10 kilometers of the project.

During the course of my meeting with I. M., we discussed this seismic risk. He proposed a trip to Tehran to visit the site. While I. M. had departed earlier, Bart and I followed more or less immediately. Apparently, Bart had other projects in Tehran.

Our opening meeting was over dinner with our client; whose mansion was impressive and the quality of the food was high. Traveling musicians and other performers provided a wealth of entertainment, but there was little appetite for serious discussion of the proposed project.

In the morning, I. M. and I headed for the site via a chauffeur-driven limousine. Traveling north on Pahlavi Road (now Valaisr Avenue), I observed a large construction site with a steeply inclined earthquake faulting clearly visible. With some difficulty we were able to accomplish a U-turn so as to inspect the site more closely. Indeed, there was a very significant fault perhaps a meter in width, with no matching of the soil stratigraphy over the exposed height of perhaps 10 meters. Also, in a neighboring construction site, the fault appeared on the opposite side of the street, perhaps indicating that this fault might cross our site. Without extraordinary measures, how could any architect or engineer, contractor or developer consider the construction of a medium-rise residential building atop that which appears to be a significant fault? Wonders will never cease!

After taking a few photographs, we continued north toward our site. Leaving the limousine behind, I. M. and I set out on foot over the moderately-steep hillside. Quite by accident, I discovered what appeared to be an animal's den, hidden amongst the brush. Peering into the same, the bottom was not visible; I dropped a stone into the hole, which stone splashed into water seemingly several seconds later. This was a very deep hole! The splash reminded me that ancient Tehran had been supplied water via underground tunnels, some as much as 100 meters below ground. Immediately, I realized that there could be many such holes and that we might fall into one of them. With no hesitation, I took I. M. by the arm, heading back for the limousine.

It was then that, out of the corner of my eye, I spotted a small pack of wild dogs. What to do? Run? Stand firm? Continue walking? Out of complete ignorance, I chose the latter, hoping that I. M. would not spot those four-legged mongrels. As the chauffeur opened the limousine door, for the first time I turned to look back. There were no dogs to be seen. Only a little dusty from our hike, we were driven back to our hotel.

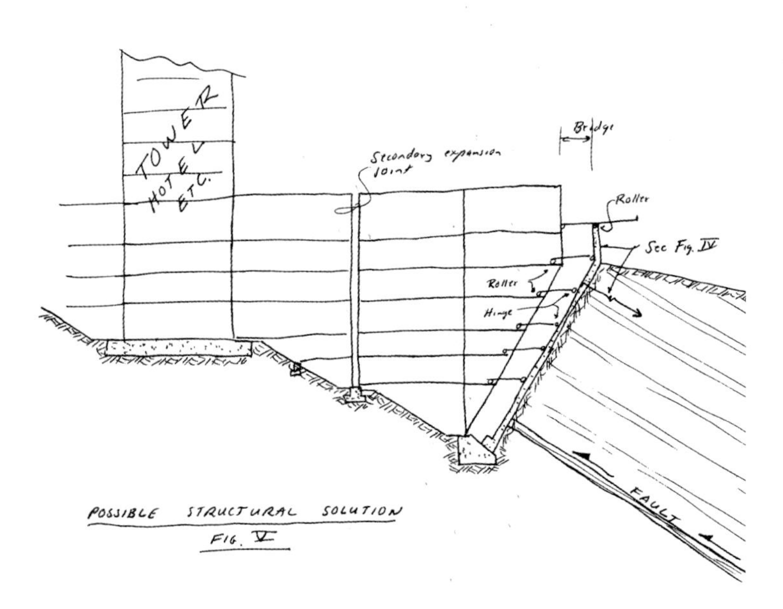

Sketch of an expensive, but effective, solution that would have allowed the construction of the Kapsad project.

We retained Woodward-Clyde Consultants, in association with Woodward-Noorany for a site evaluation. The fault was found to pass through a set of Italian-designed high-rise residential buildings. Reports by Woodward-Clyde indicated that this major east-west thrust fault bisected our site, with thrusting along the fault reaching several meters on a single event. It follows that any building constructed over the fault could experience a differential upheaval of the ground along one side of the fault; such designs, of course, are not practical.

I. M. seemed to conclude that construction was impossible. Or was it? I developed a scheme that seemed to deal safely with the construction of high-rise buildings on the site. While Woodward-Clyde reacted favorably to my proposal, the owner's cost consultant did not. Our efforts ceased, and our work went into the files. Such is life!

The really good part of this experience is that I. M. was favorably impressed with our performance on the Kapsad Project and that he and I seemed to communicate easily, leading to many further commissions and wonderful friendships with I. M., with his family, and with his staff.

The steeply inclined fault was discovered some distance from the Kapsad site but, unfortunately, it proved to pass right through the property.

Bank of China Tower

HONG KONG. 1982

Another important breakthrough in my and SawTeen's lives was the commission for the Bank of China Tower, in Hong Kong. I. M. Pei realized fully that the site was next to that of the Hong Kong and Shanghai Bank, designed by the very talented British architectural firm, Foster + Partners. The elegant HKSB was to cost three times that which was our budget for the Bank of China Tower, with essentially the same floor area but for a significantly lesser height. Despite the huge financial disparity, I. M. was not prepared to develop a design that was inferior to that of its neighbor. "Simplicity" was to be our guiding light. As well, many years earlier, I. M.'s father had been manager of the Bank of China, making the challenge even greater.

Seeking inspiration, while being a modern architect, I. M. looked to ancient Chinese philosophy and iconography. He was inspired by one Chinese proverb depicting the growth patterns of bamboo, which proverb stated that with each successive joint, a stalk of bamboo takes a measured step toward height and strength. I. M. wrote: "It is a metaphor to represent something felicitous, something ambitious, that you are always aiming high." With Hong Kong soon to revert to China from British sovereignty, the new Bank of China Tower would likely become a symbol of the presence of China, perhaps even a show of goodwill.

I. M. thought on the problem, turning it around in the idle moments (which are few) in his mind. Then one day he called, asking that I stop at his offices to discuss his proposal for "a very tall building in Hong Kong."

THE BEGINNING

Before our meeting, I learned as much as I was able about the environmental loads (wind and earthquake), as well as the foundation conditions that might be encountered. The critical information was that the codified wind loads of Hong Kong were twice those of New York, and four times larger than the codified earthquake loads of Los Angeles. Because of our prior experience with the multi-building Harbour City Project in Hong Kong, which was much lower, this research was quick and easy. It is good to recall that the lateral pressure from the wind is stronger than the occupancy loads on the floors; for a very tall building, these lateral loads from the wind will dominate the structural design. Having lived in Hong Kong, I. M. intuitively understood

This chart compares wind-induced lateral force requirements for various cities; for comparison, earthquake lateral forces for Los Angeles are included.

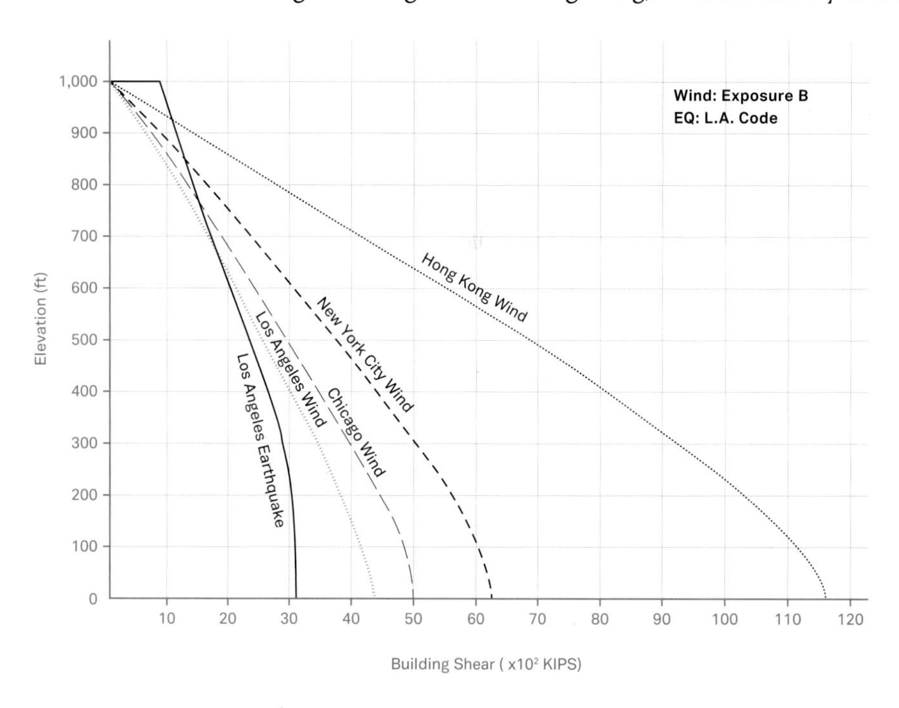

Weight of structural steel increases with the height of buildings. The least quantity of structural steel per unit of area is used for those buildings on the left in this graph.

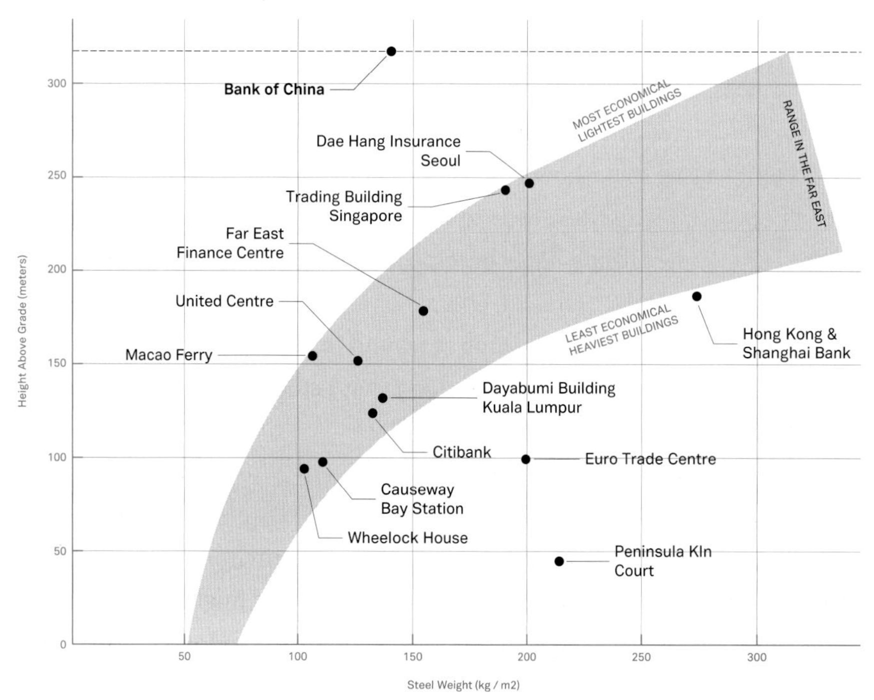

something of the ferocity of its typhoon winds and was appropriately concerned. A comparison of the lateral loads imposed on a tall tower is shown on the opposite page.

On arrival at I. M.'s office, he showed me a cardboard model of his idea: a square building, 54 meters on a side, 76 stories, and 367 meters high, to become the tallest building in the world outside of New York and Chicago. The tower was to be divided by diagonals into four identical triangles, each reaching to a different height (shown right). This put the center of the lateral forces from wind and earthquake rather far from the center of the base of the building. Further, a single central services core was not consistent with the division of the building into four identical triangles. From this, it was clear to me that the stiffness of the lateral force system need vary in accord with the centroid of the wind loads, and that this system should be located on the facade.

A sky view taken from the front of the building shows the precision of the perimeter bracing.

Almost immediately, my mind turned to the use of perimeter bracing. I. M. not only accepted this concept but was enthusiastic about it. Clearly, we had a wonderful start on an exciting adventure. But there was much to be developed in order to transfer these simple concepts into practical, low-budget systems, all inside of I. M.'s perceptive concept that "form and decoration is not enough! It must be structurally sound and elegant," but that its construction cost must be astonishingly low, with a firm budget of HK$1 billion/US$130 million (in 1983 dollars).

In turn, I. M. proposed that the facade display the bracing to be located immediately inside. This required that the bracing follow the architectural facade module, resulting in the diagonals meeting at the precise corner of the building, which intersection was highly refined in the facade design, but impossible to achieve in the structure. Experience had taught me that, with I. M., every aspect of a building has a reason, a starting point and a stopping point. Fortunately, SawTeen and I had anticipated the need to meet this "impossible" requirement. In our own meetings we had turned over ideas to make this impossible geometry possible, buildable, and economical—and uncovered a most unlikely solution.

Our approach to this dilemma was to create a three-dimensional space frame making use of two-dimensional diagonally trussed frames in structural steel. Being two-, not three-dimensional, all of the steel-to-steel connections were greatly simplified with an ensuing reduction in cost and increased speed of construction. Then, at the building corners, where the plane frames meet, where conventional construction would create those three-dimensional connections, we knitted the verticals of the frames together inside of a reinforced concrete column; we labeled it a "megacolumn."

Right An overall plan plus more detailed plans showing the configuration changes of the intersections of perimeter corner columns.

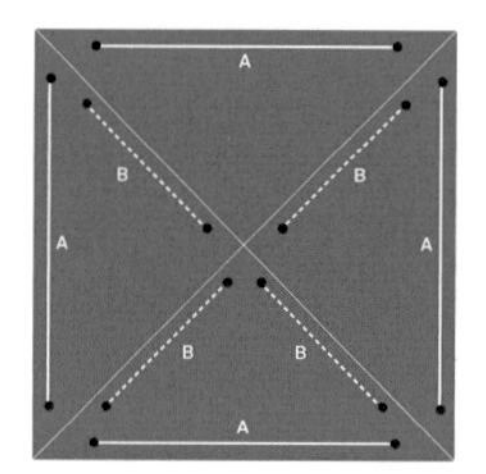

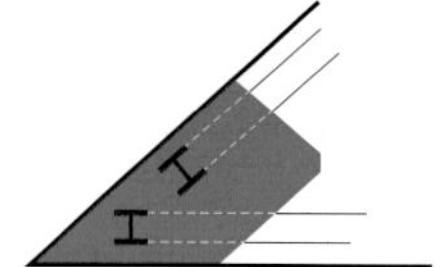

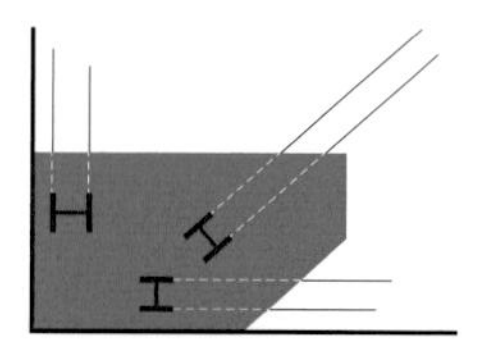

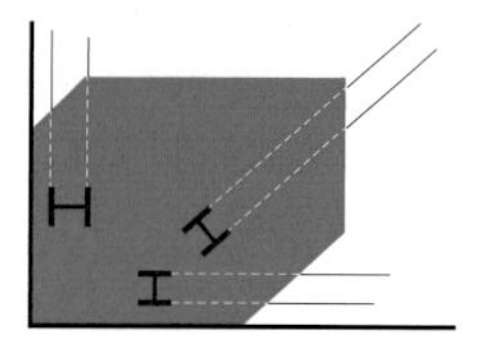

Above An elevation and plans depicting the changes in building shape with height above ground.

The potential problem with this approach is that the transfer of loads from one plane frame to its neighbors, and to the megacolumn, created eccentricities and stresses that seemed to be unacceptable. We used an approach that we later published for all to see and understand; in essence, a study in eccentricities. We were able to demonstrate that all of these eccentricities could be constructed, without creating bending moments in either the structural steel or the reinforced concrete. The structure was so constructed, standing stalwartly today while keeping us within the owner-mandated budget and speed of construction. Of interest, while the cost of the building exceeded the budget, the structure came in under its budget.

THE DESIGN CONTRACT

I travelled to Hong Kong both to select an associate engineer and to work out the details of our consulting contract. While we designed all aspects of the structures for the tower, the foundations, and the surrounding below-grade work, not being a resident, it was nearly impossible for us to obtain professional licensing in Hong Kong. Although I. M. had provided names of engineering firms who might act as the structural engineer of record, I found none of them to be appropriate. In the end, I appointed Valentine Laurie & Davies (VLD) for this assignment; ably, and under the full-time supervision of our W. John Pugh, they provided the most of the surveillance engineers.

Top The mega-diagonal under construction.

Above Pretending to hold up the lowest transfer of the mega-columns at the center of the building.

The consulting contract was more difficult. It was relatively easy to establish the scope of work and the amount of our fee for professional services. The sticking point was that the Bank of China insisted that we be paid in Hong Kong dollars, while I was equally insistent that we be paid in US dollars, but that VLD could be paid in Hong Kong dollars. Of course, I pointed out that the Bank of China, not our firm, was in the international monetary business! In the end, I drew a line in the sand, returning to New York with the knowledge that others would be the Structural Engineer. It was a very tough moment in my life! Suddenly, the Bank changed course, agreeing that US dollars would be the basis of our contract. Perhaps I. M. was instrumental in solving this dilemma. For the record, had we been paid in Hong Kong dollars, because of fluctuations in the exchange rate, we would not have been able to complete the work.

RESOLUTION OF THE ENGINEERING

Beyond the large-scale Xs on the facade expressing the bracing system, I. M.'s original concept placed Vierendeel-like trusses at what are called refuge floors: areas devoid of combustibles, where occupants are able to seek refuge in the event of fire or other emergency. I. M. had accented these trusses, visually, as broad horizontal bands across the facade. On seeing these designs, the bank's top executives in Beijing called for a meeting. Convened in Hong Kong, we were told that the Bank of China was enthusiastic about the structural system, but that they could not and would not tolerate the expression of Xs on the facade, *X* being the symbol of death in China. Many architects, unrelenting in their vision, would have thrown a temper tantrum or some such, but I. M. took the criticism in stride. We returned to New York to regroup.

Here, it would be good to share some thoughts on that ancient Chinese divination known as feng shui, touching lightly on the services of a *sifu* (master). Many persons in China, specifically much of the Chinese population of Hong Kong, take seriously the precepts of feng shui, although the Bank of China did not (at least not officially). This philosophy of harmonizing with the environment is very broad but, for purposes of brevity, this discussion will focus on the shape of the building and on the exposed X-bracing.

It would seem that the sharp 45-degree corners of the triangulated plan shape of the tower were contrary to some of the precepts of feng shui. Some of the *sifu* evidently maintained that the shape of the building was that of a "giant sword," threatening danger to neighboring buildings. This hurdle was overcome in discussions with the bank, which were all conducted in Mandarin, which I do not understand. Still, perhaps following the advice of their *sifu,* some nearby residents put chance aside, including the governor's wife, who planted a willow tree to soften views of the angular tower; others, I am told, erected mirrors outside their windows to deflect the dagger-sharp edges.

Below Shown under construction, an intersection at the center of the building of multiple mega-diagonals.

Bottom Typical corner detail of a belt truss.

BANK OF CHINA - HONG KONG BRANCH DATE: 3 June 1987. NO. BOC 458

The expression of the Xs on the facade was quite another matter. Prior to our meeting with the Bank of China executives, during an afternoon meeting in I. M.'s backyard, New York, over a glass of his fine wine, using the Golden Gate and Queensboro bridges as examples, the latter clearly visible from I. M.'s garden, we discussed a variety of architectural and structural concepts. I took the position that the expression of the heavy horizontal bands in the facade reflected an importance that far exceeded that of the structure behind, and that consideration could be given to minimizing them. The ever-perceptive I. M. pounced on that idea.

With revised architectural drawings, we returned to Hong Kong for a second meeting with the bank's Beijing executives. I. M. explained that, with the new design's softened horizontal bands, the expression on the facade was that of "diamonds," not Xs. After a short pause, the executives stated that theirs was a twenty-first-century organization, and that the expression of structure could remain on the facade. It should be noted, however, that as one

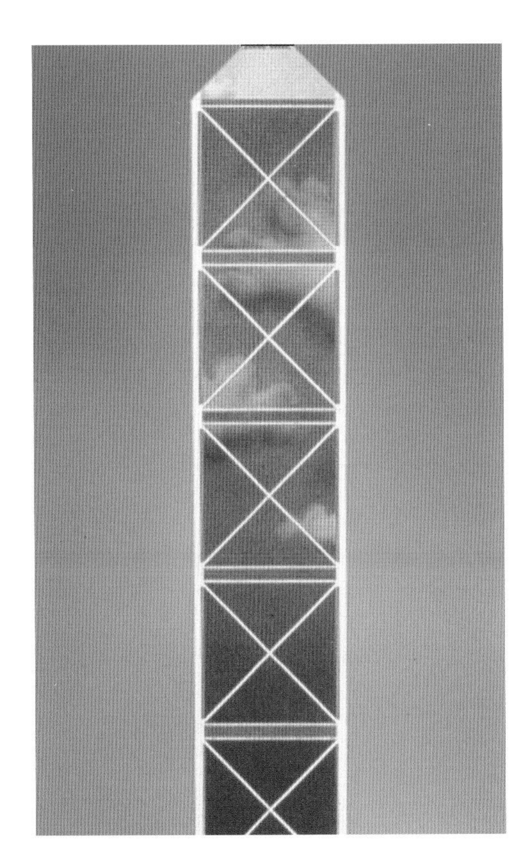

By minimizing the structural expression of heavy horizontals on the facade, the objectionable "Xs" in the initial proposal (left) were transformed into a string of "diamonds" in the revised design (right).

looks out of the building, reflected Xs are readily perceivable in the facades of surrounding buildings.

With feng shui behind us, we elected to carry the perimeter bracing sytem from the very top of the building down to the first floor above the commercial banking levels. The 54-meter clear span below was solved easily by the space frames above. Wind-induced lateral forces in the perimeter bracing are carried to the walls of the services cores through 8-millimeter-thick-steel plate floor diaphragms topped and stiffened by the overlying concrete slab. These lateral forces are then carried downward to the foundations and to the surrounding concrete slurry walls by twin, three-cell, concrete-clad shear tubes of steel plate. The shear tubes formed the elevator shafts and, at the base, the walls of the highest security bank vaults (which were concrete and re-bar clad as for a conventional vault).

THE DESIGN AND CONSTRUCTION PHASES

There were many instances where the architecture and the structure were not in harmony, but all were sensibly resolved between Bernard Rice, I. M.'s Senior Associate/Design and me. Most of these resolutions were accomplished in our weekly project meetings.

One issue was that I. M. wanted the two 54-meter high spires, reaching toward the sky from atop the building, to have a continuous taper from bottom to top.

I suggested to Bernard that a step-taper design would be faster and more economical to construct and, at that height, visually the same. Bernard reported that I. M. would much prefer the continuous taper. In a meeting with I. M., on another project, I sketched how the step-taper design allowed the various sections to be telescoped, one inside the next, and then jacked into position. This erection method eliminated the need for a tall crane atop the building. I. M.'s response was "Why don't we do that for the Bank of China?" I stated that we could and would do so.

During the erection of the spires, one of them tipped temporarily out of plumb, but of course had to be immediately straightened. Alas! I was unable to obtain a photograph of the tipped spire in this most rigorously geometric of buildings.

Of interest, at least to me, is that I am surely the only person ever to have climbed to the very top of both the non-architectural broadcasting tower of the World Trade Center in New York, 527 meters above grade, and one of the spires of the Bank of China Tower, 367 meters above grade (neither with safety equipment but both with a pounding heart).

For the most part, it would not be appropriate for me to go to I. M. with issues, leaving his staff outside of the discussion. Still, sometimes, opportunities presented themselves. For example, in the architectural drawings, the clear span of the exterior wall above Bank of China Plaza was being obscured by the detailing of the banking hall. The detailing could be seen on the back side of a large architectural model located just outside of our regular meeting area. At one point I asked I. M. to view the back of this model, resulting in I. M.'s changing the design of the banking areas to make the 54-meter (177 feet) clear span more evident. He did this, not for reasons of exposing the structural gymnastics, but because it was philosophically consistent with the design of the building; my good friend Sandi Pei, the designer of the affected banking area, has yet to forget the extra work created by this change.

As the foundations were being constructed, passing through the otherwise stalwart rock, a fault was discovered crossing one of the largest 10-meter (33-feet) diameter caissons. The time required to obtain a proper geotechnical report would have seriously delayed the work, yet all agreed that such a report was essential. What to do? While other persons were debating the issue, the contractor and I discussed a plan to excavate a portion of the fault, backfill with concrete, and continue the construction. Of course, the responsibility for this decision rested with the contractor, not with the structural engineer. Seeking to assure I. M.'s sleep, I did not belabor him of this matter. Fortunately, when completed, the geotechnical report agreed with the approach we had selected.

Bank of China Tower's topping-out ceremony was on the 8th of August, 1988, 8/8/88 being a most auspicious number for the Chinese as the lucky number "eight"

Left The construction site during the installation of the hand-dug caissons.

Below The only place where adjacent plane frames are connected with structural steel. Note the lack of fall-protection for the construction workers.

sounds very much like the Chinese word for wealth and fortune. The ceremony took place on the top floor, complete with flags, banners, speeches, and the like. The only route to the top for most of the attendees was the outside construction hoist, an experience that was not enjoyed by many. At the top of the hoist, in order to reach the floor plate of the building, one needed to step over a foot-wide gap that reached down to the base of the building. It was interesting to see people's faces as they stepped over that abyss. At the end of the ceremonies, the operators of the lift departed, reportedly stranding the caterers. Unfortunately, the only way down was via the stair, which stair had a single two-by-four wooden board at each tread but nothing in the landing. To descend, one needed to step from the tread of one flight, down and around the open landing to the top tread on the next flight. Apparently, none of the caterering staff was able to muster the courage to attempt this descent; they must have spent the night at the topping-out floor, 70 stories above Hong Kong.

At the banquet following the event, the chief iron worker presented to me a beautiful felt-lined wooden box containing two so-called golden bolts, together with a golden wrench and hammer, and white gloves. It was a magical moment for me.

POST CONSTRUCTION

The Bank of China Tower has been struck by typhoon (hurricane) force winds on many occasions and will likely experience a typhoon of Force 10 at one or more times in its life span. Because of the unique shape of the tower, and being topped with two free-standing spires well above the area of the wind influenced by the presence of the tower, there was a unique opportunity to mount wind-speed measuring anemometers at an ideal location. This tower, then, offered to us the opportunity to measure wind speeds immediately above a major city in juxtaposition to a complex waterfront and mountain terrain. Additionally the installation offered the opportunity to correlate data obtained from the spire-mounted wind sensors with similar equipment located in the near proximity of the building. Similar measurements, by wind tunnel simulation, had been accomplished for the Bank of China Tower at the Alan G. Davenport Wind Tunnel Laboratory in Canada, allowing a comparison of events in the full scale with events simulated in the laboratory. Finally, with the spire-mounted equipment, useful and necessary information could be provided to Bank of China's Building Management Office. Along with such immediate benefits, subsidiary benefits would accrue to researchers and building code officials from across the world.

Funding for the development of the program was provided by the United States National Science Foundation and by our own company, Leslie E. Robertson Associates. While the equipment was installed, for complex reasons not fully understood by me, the full benefits of the program were never realized.

I. M. gives me far more credit for the design of the Bank of China Tower than I deserve. Because of I. M.'s designs, Bank of China Tower has aged extremely well, blending softly into the architectural world and has been emulated by structural engineers of other very tall buildings. Omitting considerations of the recently added architectural lighting, in my mind's eye, it is looking even better today than when it was first constructed. In my view, it is because of I. M.'s genius and talent that this most beautiful and technically superb building will be admired by future generations. We are so appreciative that I. M., Sandi Pei, and the Bank of China allowed us to contribute to this incredible building.

The Bank of China Tower immediately became an icon of Hong Kong, appearing on postage stamps and currency.

Shinji Shumekai

SHIGA PREFECTURE, JAPAN

Tower of Angels 1988
Miho Bridge 1991
Miho Museum 1991
Miho Institute of Aesthetics Chapel 2007

In 1988, I. M. Pei received the commission to design the Tower of Angels Bell Tower for Shinji Shumeikai (Shumei), just across the plaza fronting the sanctuary that Minoru Yamasaki had completed six years earlier (see page 165). Yoshikatsu Tsuboi, the structural engineer for the sanctuary, was, of course, a natural choice for the bell tower. But the architect and the engineer, both being delightful to know and impressively talented, were highly opinionated and stubborn. Understandably, they were not always happy with their relationship.

Working on a Bell Tower for the Crystal Cathedral in California, with Philip Johnson, I had traveled to the Royal Eijsbouts Foundry in Holland to sort out the earthquake-resistant bell mounts to be used in Philip's carillon. While at the foundry, the technicians showed me the bells they were fabricating for I. M.'s bell tower, noting that it was their responsibility to design the mounts, which mounts were very different from those that I had designed for Philip. While it was not my responsibility, I sketched seismic-resistant mounts for I. M.'s bells (the technicians and I were by this time friends), making the fabricators very happy, all with little effort on my part . . . and with proper disclaimers regarding my professional responsibilities. I did not inform Tsuboi-san that I had meddled in his project. I told I. M., of course, and he thanked me, but my participation was likely little known to anyone in Japan.

By this point I had established a relatively close understanding with the Koyama family, but with no contractual agreement nor any financial compensation for my efforts. It was then that there came the passing of the world-famous Professor Tsuboi.

Opposite The elegant Miho Museum Bridge.

Below The Tower of Angels.

THE SITE FOR MIHO MUSEUM AND BRIDGE

In 1990, the leadership of Shinji Shumeikai decided to construct a museum, with I. M. being the logical choice as the architect. The next step, selecting us as the structural engineer was, to the best of my understanding, seen as appropriate by all parties.

I. M. examined many sites before settling on a forested nature preserve located atop a rugged mountain ridge. From the proposed site one could see, almost 2 kilometers away, the Bell Tower and a portion of Shumei's Sanctuary. There was yet another mountain ridge between the chosen but inaccessible site and a possible, yet to be constructed, access road. The Koyamas did not own the land but were confident that it could be purchased.

On one of our early visits to the site, Eileen Pei joined us, transported over the rough terrain in a sedan chair provided by Shumei, which arrangement she found embarrassing. As we meandered about the ridge, the adventurous I. M. elected to start down an extremely steep slope. While it was covered with small trees and shrubs, his position seemed to be precarious, causing me to rush down the slope, take him by the arm and guide him upward to safer ground; while understanding my motives, I. M. was not all that appreciative of my efforts.

There existed a strict height-above-grade limitation for the museum. But for skylights, all roofs were to be of traditional Japanese tile. I. M.'s approach to these restrictions was both brilliant and simple: remove the top of the ridge, construct the museum and then backfill to attain the maximum permitted height above grade, and use skylights for almost all of the roofs

As well, the proposed site was one ridge and one valley away from the possible access road. I. M. suggested that an access tunnel be cut through the ridge and that a 200-meter span bridge (650 feet) be constructed across the wooded valley. Additionally, a service road was an obvious requirement. I. M.'s proposal was to construct a second tunnel, sort of U-shaped in plan, with a long access road around the valley to the site. All of the above would appear to be quite expensive . . . and it was.

THE MIHO MUSEUM

We were commissioned by Shumei to produce the structural design development for the museum and the detail designs for the skylight roofs. While needing to respond to the meandering contours of the mountain ridge, the design development was quite straightforward.

The skylight structures were rather more difficult. The roof planes were established by I. M., our responsibility being limited to the development and the detailing of appropriate structural systems. Limited by the discipline of the roof geometries, I selected a space-frame concept, making use of 190-millimeter diameter

Below An overview of the site, with the bridge emerging from the mouth of the tunnel. Finishes being largely limited to traditional Japanese tiling or to skylights, much of the museum is surrounded by earth fill.

Bottom left
Excavation in progress. Note the careful preservation of the surrounding forests.

Bottom right
Copious skylights illuminate the interior of the museum, 80 percent of which is surrounded by earth embankments.

pipes. Because of the ever-changing roof geometry, the space-frame intersections were geometrically complicated. Still, it was straightforward to obtain I. M.'s acceptance of our concepts.

Standing on I. M.'s concept that the space frame node be a sphere, my idea was to make use of castings at the joint at each end of the pipes. Six bolts were used to connect the plates projecting from the pipe to the single plate projecting from the 190-millimeter (7.5-inch) spherical node. The plates were needed to make a graceful connection. The complication occurred because the erection sequence and the angle of the plates with respect to the central node needed to be properly coordinated.

We completed detailed drawings for each of the 106 different nodes. I used a reveal to divorce the casting from the pipe members of the space-frame, thus disguising the welded intersection. It was then that I discovered that, as mandated by local building regulations, each of 106 unique connections needed to be physically tested for adequate strength. Under this limitation, the reveal, hiding the welds, was not practical. What to do?

Using our friendships with Nippon Steel Company, I visited several fabrication works. There, I was told that it was possible to fabricate the castings to a strict tolerance, to press the ends of the pipe to make them more 'round' than was possible in rolling, and to then weld the two together, by which means the joint would become invisible. It seemed to require an impossibly high level of craftsmanship but, in fact, it was exquisitely achieved.

To compensate for their not inconsiderable weight, I specified that, prior to erection of the metal and glass skylights, the space frames were to be preloaded with sandbags matching the weight of the skylights, which bags were then shored to prevent further vertical displacement during the erection of the skylights. In this way, following completion of the skylight and the removal of the sandbags, the skylight remained completely flat, supported on a "deflection-free" structural system.

I. M. had proposed a sunscreen composed of wooden bars of hinoki cypress, closely-spaced and 25-millimeters square. I cautioned that these surfaces of wood represented a not inconsiderable fire hazard. With his usual gracious smile, I. M. told me that "We need not worry, the Japanese will work it out" and, indeed, they did. Visually, it is impossible to tell the difference between bars of solid wood from the as-provided bars of aluminum overlaid with a thin skin of that hinoki cypress.

A description of this lovely museum is far beyond the scope of this book. The elegance of I. M.'s design surpasses words and pictures. Where you are able, you will find a visit to the Miho Museum a wonder-filled and exciting experience.

Three views of the complex skylights atop the museum. The bottom photograph shows the sunscreen composed of square aluminum tubing, laminated with a thin hinoki cypress skin and rotated 45 degrees with respect to the plane of the glass, as well as a complex detail of a space-frame intersection.

THE MIHO MUSEUM BRIDGE

We were commissioned by Shinji Shumeikai to produce the complete design (Construction Documents) for a 7.5-meter-wide bridge, spanning 120 meters from the mouth of a tunnel to the entry plaza of the museum. While sidewalks are not defined, the principal function of the bridge is to provide an elegant path carrying pedestrians, wheelchairs, and electric carts to the museum; despite the presence of the services tunnel, the bridge needed to be designed also to support bullet-resistant automobiles transporting dignitaries, small fire trucks, and other emergency vehicles.

The dynamic behavior of the bridge, under excitation by earthquakes, typhoon winds, pedestrians, and both heavy and light vehicular traffic provided the guiding criteria for much of the structural system. Protection to small vehicles and pedestrians from down-slope winds in the relatively narrow valley, while studied, did not prove to be a problem.

Beneath, we were told, the land was sacred, the trees having been used for ancient temple construction in Nara. In my mind, it was necessary to cantilever far enough so that the bridge could be constructed from the cantilevered structure, but without supports carrying down to the valley floor. My concept was to accomplish this scheme using post-tensioned cables carried back to the tunnel thus forming the root (anchor) of the cantilever. We produced some sketches, and sent them on to Shimizu Corporation, the contractor, and the designer of the civil engineering works and structural engineers for the tunnel. We received a curt note cautioning that our design could pull a portion of the tunnel out of the mountain. I followed with a detailed description of the proposed structure, asserting that there was no unbalanced tension onto the tunnel, but that we were introducing a bending moment and a modest splitting force that need be resisted by the tunnel and by its surrounding earth and rock. With surprisingly few meetings, acceptance was given by Shimizu to pursue our concept.

Supported on 96 post-tensioned steel cables, which shimmer against the blueness of the mountain sky, the bridge seems to float without support for much of its length. For those walking over the bridge, the cables, and the steel arch that inclines forward from the tunnel seem to mysteriously vanish as they reach down to the deck of the bridge. The beauty of many other short-span bridges appears only as a view from the side, with the surface of the bridge merely a roadway. It was my intent to provide an uplifting experience for those passing over the bridge as well as to those seeing it from the side, where most people would view it. As they rise up the arch, the cables gradually increase in diameter and length with the largest being 60 millimeters in diameter and 48 meters in length. However, in order to create a visual termination of the cable grid, a few of the lowest cables are gradually increased in diameter.

Peering through the cables of the Miho Museum Bridge, a view from atop the mouth on the tunnel, with Meishusama Hall and the Tower of Angels in the far background.

Many persons have asked of the shape of the supporting arch. In truth, its shape is not in response to some theoretical force diagram but, rather, it is an arbitrary shape that I chose to be in harmony with the overall aesthetic of the bridge while providing headroom for pedestrians and light vehicles.

As with our sketches, we submitted early detail drawings to Shimizu Corporation for review. Our design had but a single 250-millimeter (10-inch) diameter pipe as the bottom chord. Shimizu replied that all was fine except for the bottom chord, which, they insisted, needed to be 1 meter in diameter—four times larger than that contained in our designs. Pushing the panic button, I put three separate teams to work, checking our designs. All concluded that our 250-millimeter pipe was just fine, and that Shimizu's comment was in error. Puzzling over this, and making my own calculations, I stumbled on the fact that the Shimizu engineers had not considered that our cables were post tensioning the bridge, constructing it low, and lifting it with the post-tension cables. This approach moved the thrust line to be nearly at the top chord where there were four times as many members, thus significantly reducing the forces in the single bottom chord. The Shimizu engineers accepted our designs.

Right The Miho Museum Bridge and entrance to the tunnel.

Bottom Hundreds of people attended the museum's opening events. The invited guests were asked to stop at the granite circle in the paving . . . and did so. This, perhaps, would be one of the heaviest loads that the bridge will experience.

Opposite A view from the mouth of the tunnel, then the arch, comprising a spectacular and exciting procession to the museum.

I structured the concrete mouth of the tunnel to follow exactly the outline of the cables of the bridge. That is, as seen from the museum end of the bridge, the concrete work is equidistant from the center of the cables; resulting in a slightly bell-shaped outline for the concrete work.

I. M.'s team designed a concrete headwall, laid back against the hillside, giving the tunnel a squared-off appearance. While I did not like the design, which was totally unnecessary, and while I argued strongly for its removal, I did not bypass staff by taking the matter into my own hands. However, one day at the site, standing with I. M. on the museum's entry plaza while looking over the span of the bridge to the tunnel, I could not help but say "Wouldn't it have been better without that squared-off concrete work?" I. M. reflected for a moment before saying "Yes, but we needed it for erosion control." My response was swift and to the point: "No, it's not needed for erosion control, I. M., it's your design." There was a brief pause before I. M. marched across the bridge to instruct the foreman to remove the offending concrete work. Perhaps it can be said that I was irresponsible in not voicing earlier to I. M. my concerns about the aesthetics and practicality of the concrete headwall. My excuse, the need to respect the designs of I. M.'s staff, is open to criticism. In a related instance, without my knowledge and for reasons not understood by me, I. M.'s staff produced and had implemented an alternative design for the handrail of the bridge, heavier than I'd intended, but beautifully executed.

One of our more unusual challenges was the removal of rainwater from the surface of the bridge. Because of erosion, it was not possible to just drain the water directly to the landscape below. For good reasons—high heels, fear of heights and the like—open grating was not acceptable to Shumei, or to me. The drain piping needed to bring the rain water to the abutment was larger in diameter than our structure! On a flight to Japan, I recalled a tennis court that I'd seen in Florida, which court was covered with a ceramic porous material, allowing play to resume immediately following the cessation of rain. "Why not," I wondered, "use stainless steel grating in-filled with that porous ceramic?"

I arranged a meeting with a high-level executive from a grating manufacturer. We did business over beer and sake before he agreed to construct a sample. On the other hand, Shimizu Corporation was understandably concerned, and not willing to experiment with the bridge decking. However, on seeing the simplicity, the beauty, and the practicality of our sample, Shimizu and Shumei readily agreed to our idea. Of course, Shimizu withheld final approval until the completion of freeze-thaw and endurance tests. Beyond the beauty of the surface, the solution offered significant subsidiary benefits:

- A free-draining roadway, similar to that of a steel grating, allowing water to pass through without creating erosion of the soils below, thus eliminating the need for drainage piping;
- Unlike the open grating that is used commonly on bridges, ceramic infill that creates a visually closed, user-friendly environment for those with vertigo;
- A dignified and interesting walking surface, fully compatible with the processional role of the bridge;
- A high-friction surface, both in wet and in dry conditions, assuring a non-slip experience for both pedestrians and electric carts, even in heavy winds;
- Without a defined walkway, a surface rough enough so that the sound of an electric vehicle approaching from behind is able to be heard by otherwise unwary pedestrians.

The floor of the approach tunnel, designed by others, was asphalt, sort of a 1930s concept, with just a little broken glass to add sparkle. Walking from the asphalt onto our ceramic bridge deck was a bit like walking from a muddy road onto a red carpet. While it was some days after the opening ceremony, I. M. solved the problem by removing the asphalt of the tunnel's roadway, replacing it with the porous ceramic for a more-or-less continuous flow onto the bridge.

The Structural Engineers Association of Illinois honored the Miho Museum Bridge with its Most Innovative Structure Award. Judge Mary B. Richardson-Lowry commented that it is "a beautiful project. The concentric arches seem to reach out to visitors and engulf them gently into the mountainside." Judge Leonard Peterson continued:

"The Miho Museum Bridge design and structural engineering integrate artistic beauty and sheer elegance. Post-tensioning finesse, small structural elements, and an innovative drainage system achieve for the designer the objective of a light and airy structure. Both Michelangelo and Leonardo da Vinci would have been in awe."

In addition to many other honors, this little bridge received from the International Association of Bridge and Structural Engineers (IABSE), its award as the outstanding structure of the year (2002). While I. M., in referring to the bridge, will almost always speak of it as "Les's bridge," I know well that he reviewed my work with care and would not have hesitated to ask for a change. Indeed, to create shadows, I. M. asked that I laminate plates onto the flat surfaces of the arch, and I did so. The very talented Paul Marantz of Fisher Marantz Renfro Stone designed the lighting.

Despite that fact that I. M. insists that the bridge is my design, not his, I know in my heart that much of what I know of design, I've learned from I. M. Unfortunately, photographs of the bridge have found their way into many publications thus detracting, in my view, from the wonders of I. M.'s designs for the museum.

THE CHAPEL FOR THE MIHO INSTITUTE OF AESTHETICS

While I. M. and I often discussed projects over lunch, for Shumei's chapel, he invited me to lunch at one of his favorite New York restaurants, Gramercy Tavern, where he announced that the Miho Chapel would be his last project, and indeed it was. Of course, on several earlier occasions I. M. had told me that a given project would be his last, and it wasn't. His continuing energy and love of architecture stood firmly in the way of retirement. Even after I. M. officially stepped away from the firm he had founded forty years earlier, he came into the office and continued to practice for another two decades.

We talked over lunch of this and that before turning to some sketches of the proposed chapel. The challenge was the structure of the roof. I. M. explained that the shape of the roof was derived from that of a traditional Japanese fan. I outlined possible materials and methods of construction before turning to my first impression. "I. M., the shape of the roof may have been derived from a Japanese fan, but it seems to me that its folded texture is closer to that of a Danish lampshade." There was a pause as I. M. considered my thought. At our next meeting, the shape continued to echo a Japanese fan, but the corrugated Danish lampshade texture was now simplified to a sleek, smooth surface, a surface that appeared again in the wooden lining of the chapel's interior.

In 2012, during opening ceremonies for Miho Chapel, an evening party was organized by the local team for those who had been active in the design and construction. Early in the party, SawTeen and I could not help but notice that I. M. had taken only a sip from his glass. Realizing that the quality of the wine was less than he would prefer, and that there was no recourse to a better vintage, we hit upon the idea of proffering a glass—no, it was two glasses, one for I. M. and one for Eileen—of Yamazaki whiskey. While the whiskey was rather more expensive than mortals could afford, we were not hosting the party, and filled the glasses rather fuller than one would normally do. Little did we realize that this would lead to a Christmas ritual wherein SawTeen and I would arrive at the Pei home at midday, toting a bottle of that delicious Japanese whiskey.

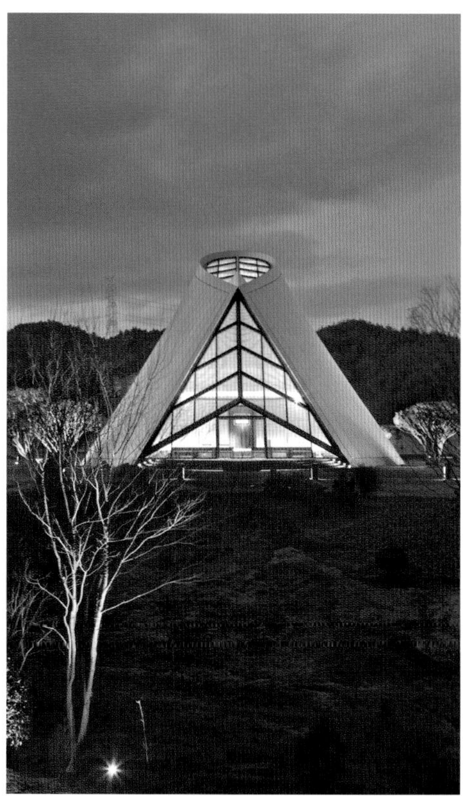

Above Addressing I. M. and Eileen Pei before the opening day of the Chapel of the Miho Institute of Aesthetics in 2012.

Far left and left Exterior and interior views of the 240-seat chapel.

Pei Partnership Architects

Chien Chung "Didi" (meaning second son, on the left of the photograph opposite) Pei and Li Chung "Sandi" (meaning third son, on the right) Pei were born to the renowned architect I. M. Pei and Eileen Loo. Both sons graduated from the Harvard Graduate School of Design, and both joined their father at I. M. Pei & Partners, there to develop wonderful designs for a variety of projects, domestic as well as international. In 1991 Didi and Sandi formed their own company, Pei Partnership Architects (PPA).

The brothers' opening of PPA was a critical step in establishing an identity independent of their illustrious father. Indeed, I. M.'s association with Didi and Sandi was both a blessing and a curse; a blessing because it offered the young architects sage advice and assisted them in obtaining some of their commissions, but a curse because PPA was sometimes misunderstood as having I. M. at the helm.

With the passage of time, I. M.'s input into PPA has continued to decline. Today, Didi and Sandi have become masters unto themselves, interpreting and putting into practice much of what they had previously learned by working with their father, augmented by their own independent approaches to architecture. This has allowed the brothers, separately and together, to generate their own philosophy, their own designs, but with an in-depth understanding of urban planning, with an appreciation of materials and technology, and with an architectural mastery not often found in architects of their generation.

Shortly after the formation of PPA, Didi and Sandi asked if I knew of an architect able to assume a responsible position in the project management of their firm. Stephen Achilles (Steve), having recently departed from the firm of Philip Johnson and John Burgee, was a logical candidate; I was appreciative of his leadership role in getting Johnson/Burgee's Puerta de Europa project in Spain both documented and constructed. Fortunately, Steve turned out to be a good fit, strengthening PPA with the technical know-how and experience required to actualize the firm's creative designs.

While Didi and Sandi were with their father's firm, we had worked with them separately on different projects that revealed their respective strengths and design abilities. The BDNI Center, in Jakarta, with two high-rise buildings, was our first real project with PPA. Subsequently, we have worked with them on many projects from a super high-rise in Dubai, UAE, intended to be the tallest building in the world, to a lovely museum in Suzhou, China.

The Suzhou Museum, discussed on the following pages, is visually beautiful, simple and uncluttered, but technically complex. Here, Sandi and Didi, holding firmly and steadily to the baton, led us through to completion. It was a delight to work with these talented and personable architects.

Suzhou Museum

SUZHOU, CHINA. 2002

The I. M. Pei family has deep roots in the beautiful ancient city of Suzhou, People's Republic of China. Indeed, today, the ancestral home and gardens, known as The Lion Forest, draws vast crowds of tourists, for the most part Chinese. Unfortunately, the size of the crowds are, to my eyes, completely out of character with the serenity of the family gardens which, while large, seem to have been designed for a half-dozen persons, not for the bull-horn led, flag-waving tourist groups that now fill the spaces. It is perhaps based in part on this circumstance that led I. M. to avoid the commission to design a new art museum for Suzhou.

Ultimately, the architectural commission for the project was awarded to brothers Didi and Sandi Pei at Pei Partnership Architects (PPA). Without regard to such contracts, fortuitously, Sandi, Didi, and I. M., working closely together, immersed themselves in creating a lovely museum and garden located not far from their ancestral home.

With the museum to be located in the very heart of Suzhou's historic district, the inevitable demolition of historically significant buildings subjected the proposed planning and design to intense local scrutiny, although in my eyes, the buildings were not architecturally significant. As well, with the museum being much larger than the surrounding historic buildings, there was considerable opposition to its proposed scale. While an outsider in the process, it seemed to me that I. M.'s in-depth participation was critical to governmental acceptance of the planning and that he contributed significantly to the design and to the excellence of the finished project.

Previously, we had acted as structural engineers for the Miho Museum in Shigaraki, Japan, and for other of I. M.'s museums, which led to our opportunity to participate in the design and most of the structural detailing for the Suzhou Museum, all while standing on the shoulders of Miho.

In part, in order to break up its larger scale, the architectural planning for the museum was structured to provide an almost meandering development of inter-connected smaller buildings and courtyards. Respecting PPA/I. M.'s desire for the intermixing of tile roofs and of skylights, under their watchful eyes, we developed an exposed structural system compatible with the room layouts. The original concept for the waterproofing system for the roof, which I had opposed vigorously, made use of open-jointed tile, not unlike systems that are sometimes used in plaza paving. While not driven by this controversy, but rather by the search for design excellence, PPA/I. M. resolved the matter by eliminating the open-joint design. At the same time, well into the project, they changed most of the tile roofs to skylights and the skylights to tile roofs. Ultimately, a grey slate was substituted for the historically prevalent and locally produced thin grey tiles. While all of this required the redesign and drafting of the vast majority of our structural work it was, in our view, for the betterment of the project.

Since the structural design for the roofs was critical to the overall visual experience, I expressed my concerns to PPA/I. M. that the craftsmanship in visually exposed structural steel attained in Japan could not be replicated in China; while listening thoughtfully to my concerns, they elected to proceed; we did our best to respect the realities of construction. As it turned out, PPA/I. M. were more in tune with the craftsmanship of China than was I. Even though not of the incredible quality achieved for the Miho Museum, only the most discerning of visitors becomes aware of the difference. Without question, at least in my mind, the beauty of the work was achieved largely through the cooperation of both PPA and of I. M.

Since our contractual assignment carried through the design development phase, Ms. Di Yaping of Suzhou Design Bureau led the development of the contract documents; it was a pleasure and an honor to work with her.

The opening of the Suzhou Museum in 2006 was a magnificent affair, with persons from all walks of life, from around the world, all enjoying the unfolding of this extraordinary work of architecture. It has been a learning and a fun experience to work with these three marvelous architects, Didi, Sandi, and I. M. Pei. SawTeen and I, and all of our staff, are justifiably proud of having participated with them in this warm and delightful museum.

Before the Suzhou Museum, our experiences with I. M.'s museums were for the most part constrained by the local topographies: mountains, harbors, forests, and the like. Here, we enjoyed an almost flat site and modest environmental loads (wind and earthquake), while using some of the technologies we developed for the Miho complex. The Suzhou Museum provided for us a whole new learning experience.

Inside and out, Suzhou Museum demonstrates the elegance and simplicity of the design by I.M. Pei and Pei Partnership.

REFLECTION

Lynn Beedle and the Council on Tall Buildings and Urban Habitat

Professor Lynn S. Beedle created the Council on Tall Buildings and Urban Habitat (CTBUH) in 1969. It began with a core group of engineers; the wonderful Fazlur Khan, structural engineer, was the initial titular leader. With time, more and more architects and other design disciplines were brought into the fold. Now, in keeping with Lynn's dream, CTBUH is dominated by architects. Early in its development, the persuasive Lynn conned me into joining the core group and much later, into being the chairman (1986–1989), a post for which I was singularly unqualified.

My term of office was so long ago as to dull the memory of any but the most significant of events. Indeed, the plural is improper as my memory focuses on but one central remembrance: Prof. Lynn S. Beedle. Lynn created the Council from nothing more than his own dreams and his own energies. He dragged us, sometimes kicking and screaming, to the far corners of the globe, thus providing a sort of technical transfer of information about tall buildings to other nations, to other universities, to other technical organizations. In a way that was typical of Lynn, he gently "bulldozed" us into creating the Monographs, that stack of books that has turned up in the offices, the libraries, and the classrooms of the world. He did his gentle prodding with both enthusiasm and grace.

The truth of the matter is that, during my term, Lynn was the real chairman of the Council. My role as the titular chairman was but to nod in admiration for that which he accomplished. Being only modestly interested in assistance from me, Lynn was both the technical and the business leader of the Council. With a staff, eager, loyal, and talented, but scant in numbers, he carried a huge burden on his shoulders. Lynn explained this to me as being the result of "financial limitations." While it's likely that these limitations were both real and serious, the essence of that burden was more likely associated with his desire to be involved in the minutest details of the operation of the Council.

With military precision, Lynn outlined nearly all that came to the fore, thus allowing us to follow in his footsteps, filling in the missing pieces (which were few, and often zero) where truly important. He collected everything in countless three-drawer filing cabinets containing papers, letters, slides, notes, and so on.

Always the perfect gentleman, Lynn Beedle was the highly organized heart and soul of the Council on Tall Buildings and Urban Habitat. His passing in 2003 was mourned by all who knew him.

Travel was the story of his life. Leaving his wife Ella and daughter Helen behind to guard the homefront, it appeared that he accepted all of the speaking invitations that came his way. And they were endless! When I wrote to Lynn, followed by a one-day visit to Lehigh University, where he taught civil engineering, I suggested on all future travel that he change his air reservations from coach to business class; he argued that it was too expensive, conceding only when I became insistent.

Lynn was a consummate communicator. Every morning, Monday through Friday, as I turned on my computer, I knew that there would be a message from Lynn. On Saturday mornings, with few exceptions, he telephoned to discuss some obscure issue. His messages were largely about the activities of the Council but strayed regularly into personal comments on family and friends. Lynn must have sent individual birthday notes to hundreds of members of the Council. On the occasion of his annual holiday, often in Bermuda, the notes continued; at the end of the holiday there was always a little recap of the wonders of his days away from Lehigh and CTBUH.

At meetings of the Council, and there were so many, in the evenings Lynn preferred to be a bit of a loner; always there was a good reason for his not joining us in revelry. It was not because others would exclude him, this was just his way. When I would chide him, he would give me a big smile and say that he had just too much on his plate.

Lynn's gentle manner and smiling countenance were always present. When times got a bit rough, particularly when he was forced to deal with my errant ways, his quick smile seemed to become broader and easier to surface.

Along the way, with the Council suffering from financial woes, a small group of us met in New York to consider possible remedies to those problems. Our conclusion was that the most important problem was our founder, Prof. Lynn S. Beedle. Nothing was too small, nothing escaped his attention. Our proposal was that Lynn remain as the head of the technical side of the Council but that another person be selected to manage its administrative/financial affairs. We drafted a carefully worded memorandum to this effect and I delivered it to him. To say that the memo went over like a lead balloon would be an understatement of his reaction. In a very unlike-Lynn manner, he exploded, making it clear that he was the boss and that he fully intended to remain the boss. Of course, be it reluctantly, we acquiesced.

Not long after our unsuccessful attempt to restructure CTBUH, remaining ever cheerful, Lynn succumbed to cancer. For reasons unknown to me, he refused medical treatment, perhaps preferring an early death to a drawn-out and painful

life. On some weekends, I would collect a person or two from the Council and drive to his home in Bethlehem, Pennsylvania. I always brought a structural or an architectural model for his inspection. We didn't stay long, he tired easily, but I hoped that our visits brought a little life into his ebbing world.

With both the World Trade Center and the Sears Tower under construction, there came the question: To what points do we measure the height of the building? The roof of the Sears Tower was a bit higher than that of the World Trade Center, while the top of the antenna of World Trade Center was higher than the antenna atop the Sears Tower. In the hopes of establishing a sensible definition of the tallness of buildings, Fazlur Khan and I set up a private meeting. We agreed that the "official" height of buildings should be to the top of the architecture, excluding transmission towers and their ilk. The Sears Tower, in the eyes of the Council, became the tallest building in the world.

Little did we realize that this "decision" subsequently adopted by the Council on Tall Buildings and Urban Habitat, would lead to the construction atop buildings of countless not-so-sensible "architectural" spires. For one of our projects, the owner-developer asked me to add 2 meters to the top of an existing architectural spire, thus making the building "taller" than one of its later-constructed neighbors. Reasoning with our clients, I was able to convince them that the proposed 2-meter addition was not consistent with the dignity of their building. The addition was not constructed.

Today, the Council on Tall Buildings and Urban Habitat, has grown in size and scope beyond what we had imagined. Prof. Lynn S. Beedle created a solid foundation for the Council; standing on his shoulders, CTBUH has become a leading source of information on the high-rise building.

In 2003, at an awards dinner, several of us were asked to provide remembrances of Lynn. Unfortunately, I was called overseas, sending my thoughts in writing:

Dear Lynn:

On this evening, a prestigious evening honoring you and that which you have created, I'm to be found en route to the Far East. This letter, then, must be a poor substitute for my spoken words.

Surely you will recall that it was more than twenty years ago that you conned me into becoming the Vice Chairman of the Council on Tall Buildings and Urban Habitat...later to become a very poor Chairman.

Wasn't the big international congress of the Council held at Lehigh a bit more than thirty years ago? And the collaborations go on...the Code Advisory Committee of the American Institute of Steel Construction and so many others.

In all of these activities, you were the one to carry the heavy load, with the rest of us, burdened with only light packs, following you up the mountain.

Lynn, with this letter it's not possible to even begin to examine those activities wherein we have overlapped. Were an attempt made to examine tonight the remainder of your contributions to our profession and to society, the night would never end. Instead, this is a time to marvel at the depths and the breadth of your professional and your nonprofessional contributions to peoples, worldwide, and to respond to the warmth of our friendship.

Tonight, joining the thousands who have felt the touch of your hand, we send to you and to those you love all of our very best wishes for the years ahead.

Your devoted servant and admirer,

Leslie E. Robertson

And it is all so very true. Prof. Lynn S. Beedle, unmistakably the real chairman of the Council, was my ideal of the perfect gentleman. Would that I, as well as others, could rise to his level.

Kohn Pedersen Fox Associates

Emerging from the architectural firms of John Carl Warnecke & Associates and I. M. Pei & Partners, the remarkable triumvirate of A. Eugene Kohn (Gene, on left in the photograph opposite), William Pedersen (Bill, on the right), and Sheldon Fox, in 1976, formed Kohn Pedersen Fox Associates, which would become one of the leading architectural firms of our time. With strong leadership in design, marketing, and management, seemingly divided into those clear areas of responsibility among the partners, respectively Bill, Gene, and Sheldon, the firm grew beyond its headquarters in New York, to establish offices worldwide, including in London, Shanghai, Hong Kong, Seoul, and Abu Dhabi.

While he was heading the offices of John Carl Warnecke, New York, I had a short but happy encounter with Gene regarding a project on Wall Street. Many years were to pass before 1995, or thereabouts, when we became involved with the Shanghai World Financial Center, as described in the following pages. It was in this era that we got to know the designers Bill Pedersen, Paul Katz, Jamie von Klemperer, Bill Louie, David Malott, and so many other talented persons in the firm; unfortunately, Sheldon Fox, whom I had never met, had succumbed to brain cancer.

It was astonishing to have such imaginative architects come to us to say that a proposed project demanded a strong input from the structural engineer, .asking that we provide a structural form around which they could design a responsive architecture. Equally interesting was the opposite scenario where they would place an architectural form on the table, asking how we would provide the required structure.

The bottom line of all of this is that KPF seemed to be without hero-worship, with the young architects encouraged to provide strong input into the overall process, thus bringing to us a deeper engagement and just plain fun.

Shanghai World Financial Center

SHANGHAI, CHINA. 2001

Strategically located in the heart of Pudong's Lujiazui district, an area that has emerged as China's commercial and financial capital, the 492-meter Shanghai World Financial Center (SWFC), is an icon of Shanghai. Constructed of a composite steel and concrete megastructure, this twenty-first-century vertical city is a tribute to two brilliant and imaginative men: Minoru Mori of Mori Building Company, the developer, and Bill Pedersen of KPF, the lead architectural designer.

The tower's basic form is that of a square prism, 58 meters on a side, intersected by two sweeping arcs to form an ever-changing six-sided shape in plan, ultimately terminating in a single horizontal at the apex.

The building is mixed-use, with meeting and retail spaces at the base, a 300-room luxury hotel at the top, and 72 office floors between. Above the hotel, at the 94th to 101st levels, there is a visitor's center and observatory. Much of the three floors below grade is devoted to mechanized parking.

A BIT OF HISTORY

We had established a long relationship with Nippon Steel Company dating back to the early days of the World Trade Center. I had met and become good friends with the architect Shoske Watanabe, who had invited me to Japan to spend two weeks learning something of the beauty of the country and of its people. Shoske-san's father, Watanabe Gisuke, was chairman of Yawata Steel Company, now Nippon Steel Company. As an example of our relationship with them, as is common in the Far East, Nippon Steel commissioned us to provide an alternative design for a new project. Our agreement called for equal monthly payments. The project moved forward rather more slowly than had been anticipated so that payments to us began to exceed the designs that we had produced. I wrote to Isao Kimura of Nippon Steel suggesting that, for a time, he should stop sending the monthly payments. He replied "sometimes we believe

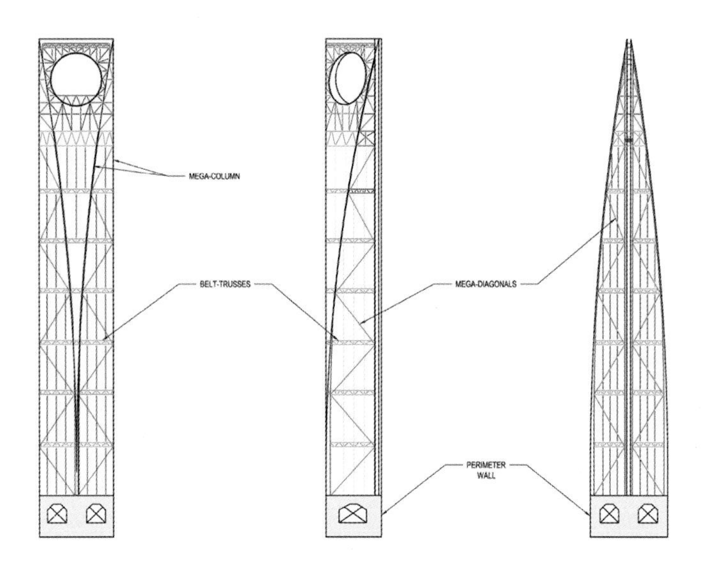

that you charge too little and sometimes we believe that you charge too much," and they continued to send the agreed-upon monthly payments.

The original designs for Shanghai World Financial Center began in 1993, with development by Mori Building Company, with KPF as the design architect, and with structural engineering by Ove Arup & Partners, New York (Arup). All engineering and the detailed architectural design but for the facade and the public spaces, moved to Tokyo, to be completed by Shimizu Corporation (Shimizu).

By 1995, the piling had been tendered and installed, and the structural package had been completed. In preparation for tendering of structural steel, we were approached by Nippon Steel with the goal of providing a lower-cost, faster-to-construct structural system. Our preliminary designs were completed in sufficient detail for tendering and submitted to Nippon Steel, where they were priced before forwarding to Mori Building Company; at that moment, perhaps stimulated by the Asian financial crisis, Mori Building Company stopped the project.

Our designs departed heavily from those of the original structural engineers, making use of a laterally stiffer perimeter frame and a lighter, less stiff, concrete services core. These changes allowed for the construction of a taller, heavier building onto foundations for the earlier structural design (by others).

In early 1998, a further query from Nippon Steel Company's Isao Kimura asked of our interest in preparing for them a complete alternative structural design for Shanghai World Financial Center; we responded in the affirmative, receiving from them architectural and structural drawings. The architectural drawings by KPF depicted an extraordinarily elegant and buildable design. The structural drawings, by Arup and Shimizu, provided for a reinforced concrete services core with a perimeter rigid frame, the latter not unlike that used in the Sears Tower, in Chicago, certainly all well within the design and construction technologies.

SawTeen, as Partner-In-Charge and Project Director, and I as Project Designer developed a proposal replacing the wide, closely spaced, perimeter columns and deep perimeter beams of the contract documents with perimeter bracing. Our proposal provided significantly enhanced views from inside the building, while reducing the construction cost and speeding the completion of the building.

It was immediately apparent to me that we needed Bill Pedersen's acceptance of our proposed structural system. I met with him in KPF's offices so as to disclose our intent. Bill not only accepted our ideas, he was enthusiastic about them! It seemed that our only obstacle was to gain acceptance by Mr. Mori.

Several months later, in August 1998, Nippon Steel informed us that Mori had terminated its contract with Shimizu and had decided to suspend the project for one year (presumably because of the continuing Asian financial crisis).

By November, we submitted five alternative basic schemes for the structural systems, with the differences largely being in the geometry of the perimeter bracing. Quick sketches were sent to Nippon Steel, there being warmly received. We then proceeded to prepare drawings suitable for tendering so that Nippon could prepare proposals for both the conforming and our alternative tender. In truth, I was supremely confident that our proposal would be accepted and that we would be sent off on yet another wonderful adventure.

In 1999, with KPF remaining as Design Architect, and with the foundation piling in place, the project was restarted but with the height of the building increased from 460 meters to 492 meters, and the base dimension increased from 55.8 meters to 58.0 meters. This increased the gross area of the building by approximately 15 percent and the overturning moment from the wind by approximately 25 percent. The exterior appearance of the proposed building remained essentially unchanged.

Under the footprint of the tower, about 200 concrete-filled steel pipe friction piles at minimum spacing had been driven by Shimizu from the ground surface. Pile cut-off was at the anticipated bottom elevation of the mat (-17.5 meters), to be used later for top-down construction. Temporary vertical support for the below-grade concrete floors was provided by steel H-piles extended from some of the piling to the ground surface.

With the rebirth of the project Mr. Mori, recalling our earlier designs, approached us seeking an alternative design to that contained in the original tender documents. In part because the pile cutoff was well below grade and in part for non-engineering considerations, the cost of reinforcing the existing piling was high. We were told that, while the mayor of Shanghai would accept the concept of the larger building, revision to the piling would not be permitted. As well, we were told that structural engineers from Japan had stated that the taller, larger building could not be constructed without modifying the existing piling, thus seemingly rejecting Mori Building Company's proposal for a taller, larger building.

With the assistance of Christopher Robertson of Shannon & Wilson, Inc. geotechnical engineers, we determined that the installed piling was capable of sustaining safely the original design loads. Accordingly, the piling could accept a larger building, but only where the weight of the new and larger building was not more than 90 percent of that of the smaller building and by redistributing the loads to the piling so as to accept the increased overturning moments from wind and earthquake. In short, we were able to achieve a reliable and constructible design for the larger building, but only where we could make the larger building lighter than the smaller building. Mr. Mori, recalling our alternative designs of 1995 and 1998, asked that I travel to Tokyo to discuss the matter.

THE NEW STRUCTURAL SYSTEM

Before venturing into such unknown territory, SawTeen and I made a quick evaluation of the proposed larger building on the existing foundations. We found that, by stiffening the perimeter bracing well above that of the original moment-resisting space frame, we could decrease the concrete of the services core to make the larger building lighter than that of the original design! Clearly, with enhanced wind engineering, a road to success was beginning to open before us. Exultant, we prepared presentation material to be viewed by Mori-san and his staff.

Accordingly, abandoning the original design for the perimeter framing (that of a Vierendeel moment-resisting space frame similar to that of the Sears Tower), the idea was to resurrect our design that Mori-san recalled from 1995: a diagonal-braced frame with added outrigger trusses. I took a flight to Tokyo.

There was a huge meeting in Mori Building Company's offices in Tokyo, with just room enough for Mori-san, his staff, Japanese architects and engineers, and me. My presentation, while in English, seemed to be well received. A follow-up technical meeting was called with just architects and engineers. Shortly into that meeting I was called away to meet with several of the lead persons from Mori Building Company. Almost immediately they asked: "Do you have other ideas?" Completely caught off guard, my response was that we had plenty of ideas, that I would return to New York to further develop those ideas. Their response was (paraphrasing): "Since you don't have additional ideas, there is no need for you to return to Tokyo. Please send us a bill for your expenses." While stunned, there was no room for discussion. Departing the Mori offices, dejected and unhappy, I took the first plane back to New York. In truth, this had been a hard pill to swallow; lots of really good ideas, ideas that had not been discovered by the very competent engineers of Japan, all leading us to nowhere.

A few days later, to our surprise and delight, Mori-san wrote, asking that I again travel to Tokyo to discuss our designs. Again, Japanese architects and engineers were present. Our presentation was able to assure them that we could produce a design for the larger and taller building that would rest safely on the existing piling. Life was good, and I returned to New York.

We worked closely with KPF, and Mori Building Company to incorporate this new structural system into the existing architectural form. Beyond the addition of bracing, by making use of outrigger trusses coupled to the columns of the megastructure, a further reduction was realized.

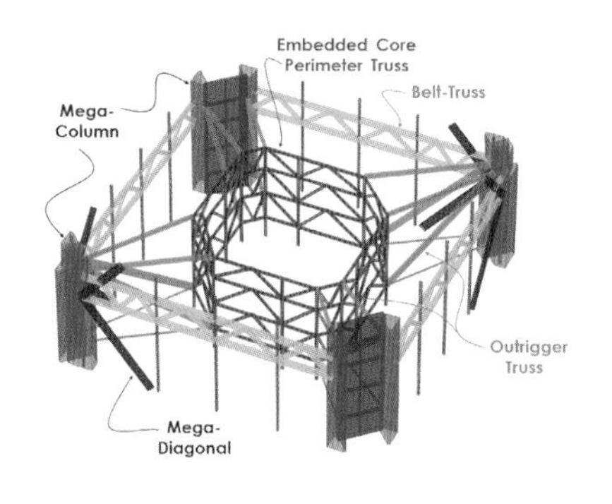

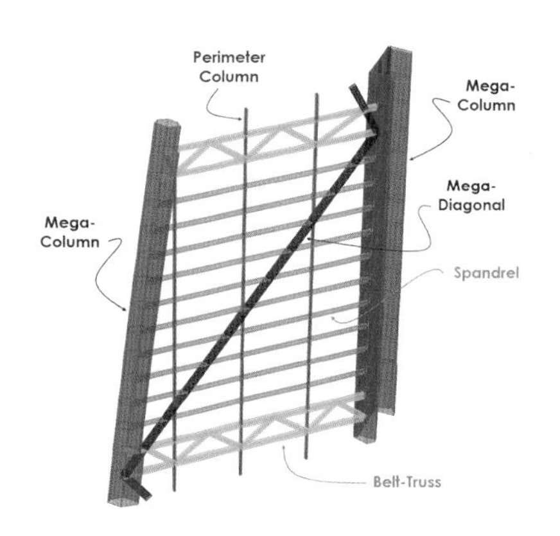

Far left In part to resist lateral forces, vertically spaced outrigger trusses were added, connecting the services core to the perimeter trusswork.

Left Here, unlike the Bank of China Tower, because of planning changes, the structural diagonals are not fully coordinated with the architectural module or with the floor lines.

A message arrived from a friend in Mori Building Company, with a copy of an untrue and damaging letter from the chief structural engineer of a Japanese architecture/engineering firm, the letter having been sent to Mori Building Company regarding our services. In the United States such a letter would likely be grounds for bringing that engineer before the Professional Board of Inquiries for disbarment but, in Japan, it must be quite different. There was nothing that we could do but to just bury the letter, hoping that it had not been favorably received by Mr. Mori. We learned later that there was a design competition between us and the same Japanese structural engineers; this was a one-way contest in that they had attended our presentations without our being informed. We knew nothing of a competition nor of anything about their proposals.

To take place in the early afternoon, yet another meeting was called by Mori-san. An agenda was not provided. On my arrival in Japan, the competing Japanese chief structural engineer proffered an invitation to lunch at an elegant restaurant with excellent Western cuisine. Over coffee, he said that they wanted to cooperate with us. I realized immediately that a decision had been made about our participation in the project. I pointed out to him that it took two, not one, to cooperate, that we were cognizant of the letter that he had sent to Mori Building Company. I asked why he had sent such a hurtful and baseless letter. Without hesitation, he said: “We wanted to win.” Astonished and stunned by his response, I shook my head; somehow, without a decision, we finished our coffee and walked to Mori Building Company’s offices. Later, we were to become friends.

The meeting was short with Mori-san standing before the same big group. Pointing to me, he said: “You shall be the structural engineer for the Shanghai World Financial Center.” Pointing to the chief structural engineer, he said: “You shall assist.” He did not define the nature of the “assistance.”

Following a return to New York, as we were completing our concept designs, the Japanese architect/engineering company being strong, competent, and proud, resigned from the project. Irie Miyake Architects and Engineers (IMAE), Mori’s in-house architectural/engineering group, but for the facade and the public spaces, completed the basic architectural tender documents. For reasons not understood by me, neither Shimizu (the original general contractors) nor Nippon Steel continued forward on the project.

THE MEGASTRUCTURE

To resist the forces from typhoon (hurricane) winds and earthquakes, SawTeen and I introduced three parallel and interacting structural systems:

1. The megastructure, consisting of the major structural columns, the diagonals, and the belt trusses.
2. The concrete shear walls of the services core.
3. As created by the outrigger trusses, the interaction between these concrete walls and the megacolumns.

KPF was able to capitalize on the presence of the outrigger trusses by beautifully incorporating them into the architectural design of the skylobby and hotel floors. Seeking to improve the quality of the office spaces on each of the four orthogonal faces, the new structural system decreased the perimeter framing from the seventeen wide columns of the moment-resisting frame to a maximum of just three narrow columns. Depending on the breadth, for the two sloping faces, one or two narrow columns were provided, or in some cases no columns at all. Hence, building occupants are provided an extraordinary sense of openness and unparalleled views of Shanghai.

The megastructure is displayed subtly behind the facade of the building. Architecturally founded on a heavy stone base, the megastructure gives the impression of both strength and of permanence. Indeed, it was one of the goals of both KPF and Mori Building Company to communicate these two attributes while retaining the wonderful elegance of the architectural form.

TECHNICAL REVIEWS

As is customary in China, for more significant buildings, technical review boards are assembled to consider critical aspects of proposed designs. For Shanghai World Financial Center, SawTeen and I called on our Wing-Pin Kwan (Winnie) and on outside consultants: for wind, Prof. Alan G. Davenport and Prof. Nicholas Isyumov, both from the University of Western Ontario, and Prof. Arthur N.L. Chiu of the University of Hawaii; for earthquake, Prof. Stephen A. Mahin from the University of California, Berkeley, and Prof. Paul C. Jennings from the California Institute of Technology; for foundations and geotechnical, Christopher A. Robertson, from Shannon & Wilson, Inc. of Seattle.

There is the tendency among experts to "talk down" to attendees at such meetings. In China, this is unwise, an insult not easily forgiven by resident experts. Despite my prior warnings on this point, mistakes were made by some outside consultants.

The discussions were carried out in Mandarin, with our Winnie translating. Winnie was from Hong Kong and could speak the Hong Kong and Mandarin dialects; to conduct a private discussion, the assembled review board simply shifted to the Shanghai dialect.

The earthquake experts seemed to be convinced that we should use crossed-bracing in lieu of the single-direction bracing that SawTeen and I had proposed. As well, they wanted bracing on the triangulated curving faces, whereas we wanted to leave those faces without bracing. In response to their concerns, we prepared detailed calculations and graphics, plus models to support our position. In the final meeting with these specialists, I made an impassioned appeal for the single-direction bracing, arguing that Shanghai World Financial Center would likely become a new symbol for Shanghai and that, when supported by sound engineering, aesthetics would figure as a key factor. Suddenly, one of the Mori Building Company's senior engineering executives stood up and told me that I must sit down. Stunned, I did so. Although other matters were yet to be discussed, the Chinese immediately closed the meeting, there being no easy recovery from this rebuke. But, why were the Mori Building Company executives so abrupt and rude? Perhaps they preferred the crossed diagonals that I had argued against? Perhaps they wanted the diagonals on the curving faces? Perhaps, understandably, they would have preferred resident, rather than imported, engineers.

What to do? These were the same executives who had earlier said to me: "Since you don't have additional ideas, there is no need for you to return to Tokyo." In truth, I fully appreciated their desire to have Japanese engineers instead of us. I am not one to take offense or to become angry but, still, I could not continue on the project under their limitations. On returning to New York, I wrote to Mori-san stating that I had resigned from the project and that others from our firm would assume my responsibilities. Mori-san wrote back to say that he appreciated my position but, sincerely, he and others in Mori Building Company wanted me to stay. Of course, I did so.

The conclusion of all of this was that the Chinese seismic experts agreed with our proposed designs without any changes in the configuration or location of the bracing.

As is customary in China, a Chinese design bureau was assigned to the project. East China Architectural Design & Research Institute Co. Ltd. became the architect and the engineer of record, with Mori Building Company providing detailed architectural assistance.

CONCLUDING THOUGHTS

The Shanghai World Financial Center is the fruit of the creativity and passion of many individuals, all with the common goal of providing a place in which people can meet and mingle, share ideas, and give birth to new knowledge, culture, and values. It provides urban spaces and environments in which the future will be created. We went on to assist KPF on many wonderful projects, both domestically and worldwide.

Special thanks need be given to Mr. Minoru Mori, who provided the vision for the design and construction; to William Pedersen, who conceived this wonderful building; and to David Malott, who directed much of the design. KPF, SawTeen, and many talented and resourceful men and women of our company took hold of this project and made it happen.

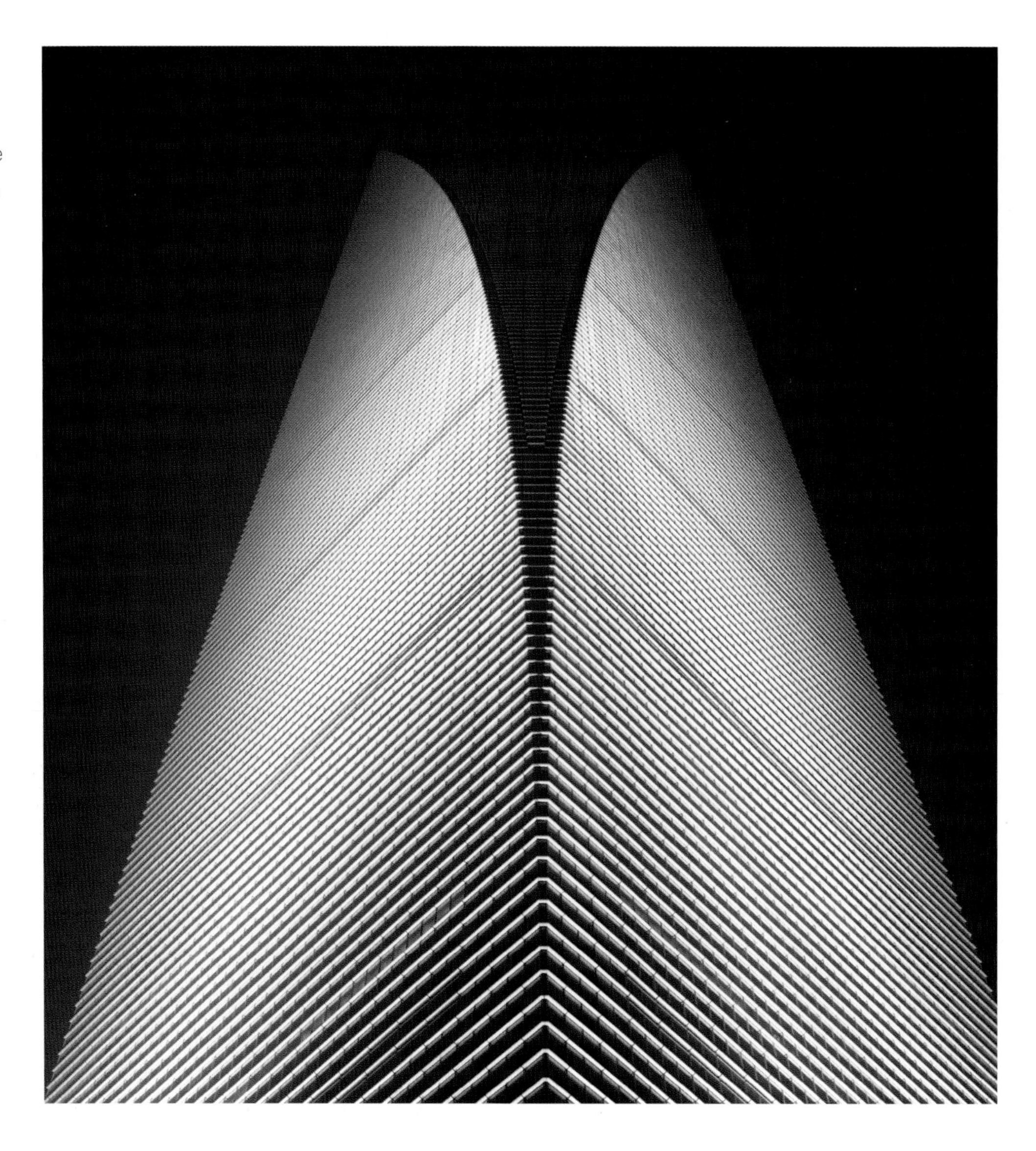

The beautiful design by KPF is accented in this photograph of an elevation of one of the two opposite corners.

Left and below left The two-dimensional braced frames were structurally integrated by the surrounding columns of the mega-structure to become a three-dimensional space frame.

Below The structural system immediately behind the architectural facade.

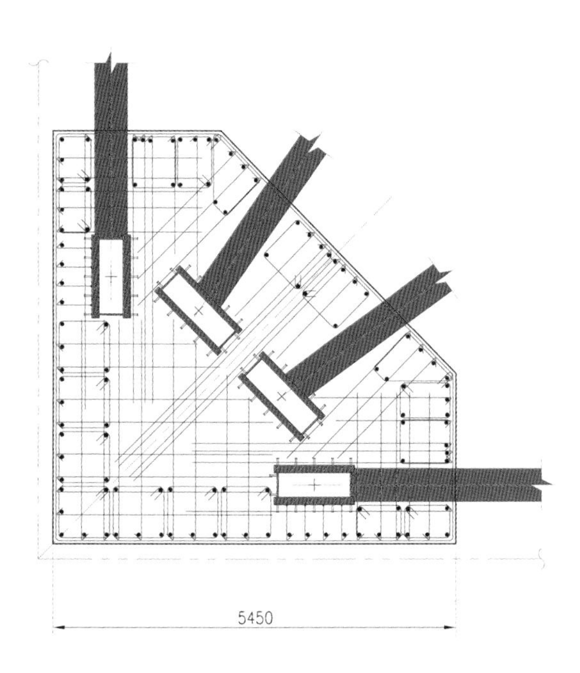

NHK Sky Tower

TOKYO, JAPAN. 2014

In January of 2014, Kohn Pedersen Fox Associates was contacted by NHK, the Japanese TV broadcaster, which sought material describing the future of building construction. KPF's David Iwami Malott, who speaks fluent Japanese, headed the team. I had worked with David in the past; despite my inability to speak Japanese (though I am able to find my way to the loo in many languages), I was chosen as the structural consultant but with SawTeen and her company participating in the design and providing the technical staff.

The larger goals of this theoretical study were aimed at some of the critical environmental and societal issues facing Tokyo and its surrounds: rising sea levels, tsunami- and wind-driven waves, waterfront development, renewable energy, transportation, and the like. The goal of the study was to develop material that would be informative and of interest to a non-technical TV audience. David proposed that we focus on a tower, 1600 meters in height (one mile), displaying something of the earthquake-resistance, robustness, and reliability, efficiency, and sustainability of the next generation of super-tall buildings. On several occasions NHK brought their film crew to New York to document some of the efforts of the design team.

David and his staff developed drawings depicting the shape of the proposed tower but, of more importance, a small wooden maquette. From this maquette, without altering David's basic design, I was able to sketch a structure of unusual simplicity and intrinsic strength. At any given level, three separate buildings are combined to form a single building with the contiguous floors in the buildings immediately above and below being rotated into the position of the vertical slots. My proposal was to provide continuous trusswork from the foundations to the very top of the building but

While other sites were studied, this site offered the least disruption to existing communities while allowing additional tsunami protection to waterfront facilities within Tokyo Bay.

with that trusswork located on the back of the occupied spaces, making it unseen (not in the plane of the glass facade). Since the occupied spaces rotates every 80 floors or so, the bracing rotates as well, leaving only half of a given level containing the trusswork.

Realizing that the wind loads on such a tall building would exceed both the earthquake loads as well as the loads imposed on the floors—the long vertical slots separating the building into thirds—was invaluable in reducing wind loads and in providing desirable aerodynamic behavior. Having worked with Pei Partnership Architects on a three-slot design and subsequently with Woods Bagot on a four-slot concept, we had come to appreciate that, in the wind, both the total wind load and the dynamic lateral response of a building composed of the three-slot design is superior to that of a four-slot concept.

In short, David's original concept was an asset in providing an avenue for the development of a superior structural system. Catching the eye of the worldwide press, the concept has spread far beyond NHK.

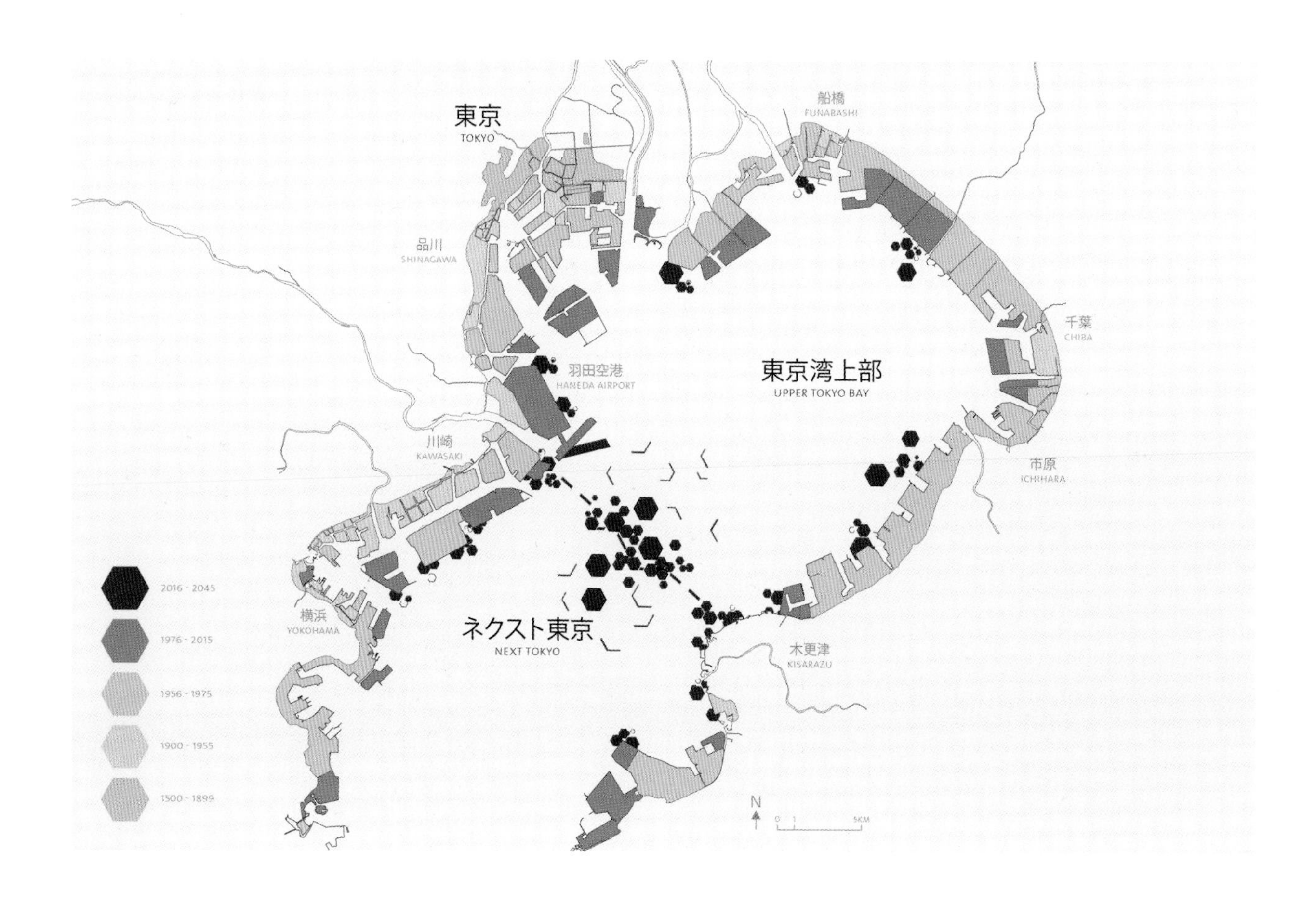

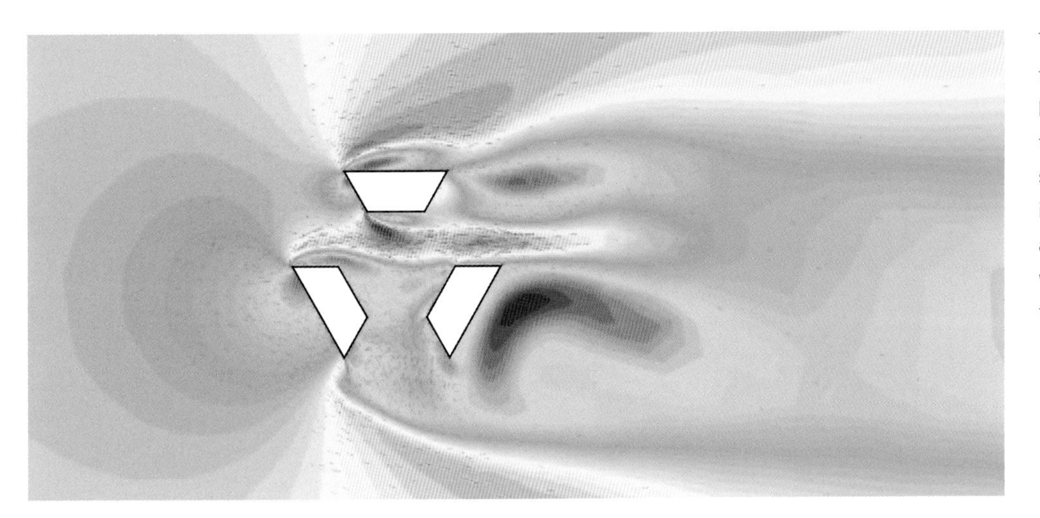

The structural concept for this proposed mile-high building allows wind to flow through a three-slot system, thus reducing both the dynamic and the steady-state wind-induced forces on the building.

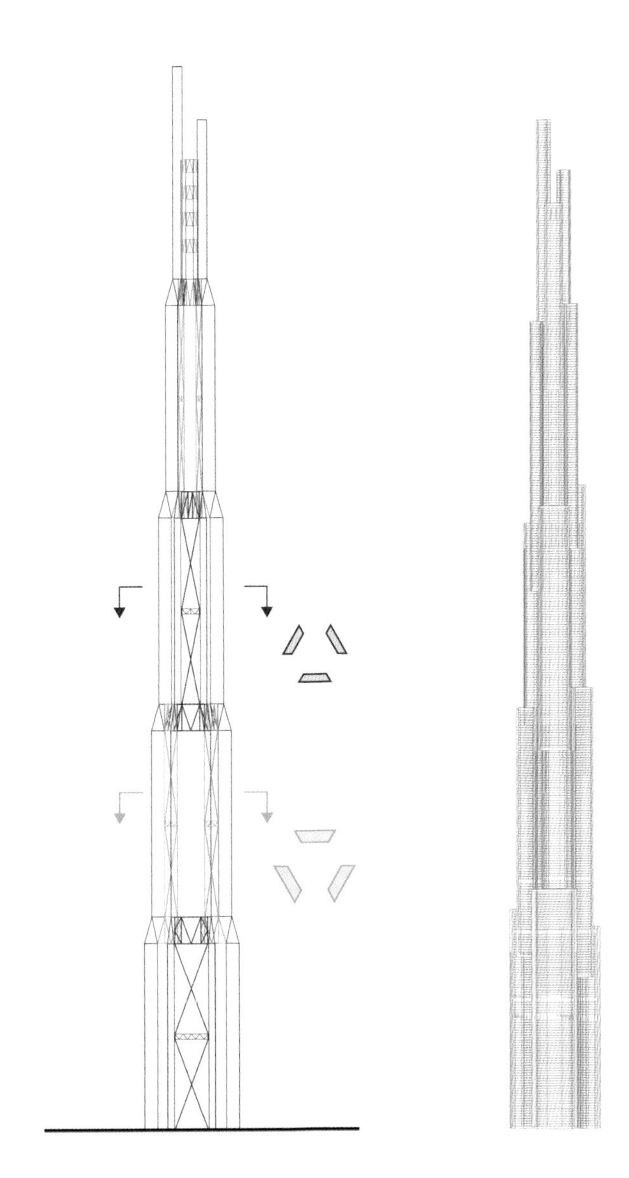

For an engineer, the opportunity to exchange ideas with sculptors and other artists is an experience that should not be missed. Shown here, the world-renowned sculptor Beverly Pepper exchanging ideas for one of her works.

Furniture, Sculpture, and Parks

While it may seem odd that a structural engineer would become involved with such things, I have always welcomed the opportunity to participate in very small projects. As well, for a young project manager, there is the opportunity for him or her to undertake the entire project—sketching, calculations, and CAD—while, at the same time, meeting and dealing with a creative mind rather out of their norm. In a sense, the design of furniture and sculpture provide the perfect stepping stone for a young architect or engineer.

FURNITURE AND LIGHT FIXTURES

Beyond the making of furniture for our home, my experience is not large. Let me provide just one example:

Working with Aldo Giurgola, I started by performing routine tasks such as the providing of minimum sizes for both metal and wooden elements. From this, there grew a participation in component shapes and the like, all leading to connection designs. To me, it was fascinating that this world-renowned architect had the interest, the time, and the skills to create beautiful pieces for home, office, or church, and allow me to assist on such projects.

While I've never worked with Bill Pedersen in his role as furniture or light fixture designer, his achievements are well known. Over lunch, Bill showed me some drawings of a chair under development. It seems that his lovely wife, Elizabeth, had counseled him that the setting out to design a new chair was fine but, unlike his last chair, this one should be affordable to mortals. Hopefully, one day, there will come the opportunity to participate with Bill on such projects.

Miscellaneous fixtures provide another example of the participation of the structural engineer in design, particularly for large overhead light fixtures and for

A massive light fixture by the very talented and imaginative I. M. Pei (with Paul Marantz of Fisher Marantz Stone and Aslihan Demirtas, consultants), installed in the Museum of Islamic Art, Doha.

theater equipment. Where the support is straightforward, say three wires or cables, unless very large, there is no need for consultation but for more complex installations, whether or not it is taken, there arises the need for advice from a structural engineer.

Without input from Philip Johnson, John Burgee added planters and light fixtures to the atrium annex of our AT&T headquarters building. The planters were quite straightforward, being just boxes founded atop the paving, but the light fixtures were hung from the arching steel skylight that I'd worked out with Philip, providing the curvaceous openings through the webs. The making of an elegant connection of the light fixture hanger to the steel arches and the structure of the light fixtures themselves proved elusive. I'm convinced that better designs could have been found but remain equally convinced that I didn't find them.

SCULPTURE

There are those who argue that all architects are frustrated sculptors and/or that all sculptors are frustrated architects. Throw engineers into the mix and you get something else.

The structural engineer is drawn into the design of sculpture for a variety of reasons, only some of which appear below:

- Review and certification as required by building authorities
- The fragility of older sculpture, particularly as impacted by seismic events

- Foundations
- Gravity, environmental and earthquake loads imposed by the sculpture onto existing construction
- Internal or external connections
- Rain, snow, and ice
- Freezing and weathering
- Thermal expansion and contraction
- Earthquake and wind
- Technical problems: bearings, metallurgy and the like
- Deflections and/or stiffness

While all of the above may come across as dull and uninteresting, in my experience, the overall experience is exciting, challenging, and fun.

SOME EXAMPLES WITH RICHARD SERRA

Richard Serra is, in my view, an exceptionally talented sculptor and painter. He has very strong views on anything that impacts his work, and is not reticent to state them. It follows that there is often a level of finality to his decisions, a fortunate circumstance in that, in sculpture, the meddling by architects, engineers, contractors and their ilk seldom leads to an improvement in the art.

For one of his undulating, wall-like forms, Richard had retained German structural engineers to provide evidence of seismic resistance. While the sculpture had been sold to an artlover to the east of San Francisco Bay, Richard sought to temporarily exhibit it in the San Francisco Museum of Modern Art. He had submitted the drawings, calculations, and the like prepared by the German engineers to the museum which, in turn, provided the same to its structural engineer. After careful review, along with the preparation of appropriate calculations, the structural engineer rejected the submission on the basis of inadequate seismic resistance.

Richard sent the original calculations to me, asking for my comments. These are very heavy pieces, with the steelwork often 4 to 6 inches thick. I agreed with the comments of the museum's structural engineer, suggesting to Richard the introduction of base isolation details to partially isolate the sculpture from earthquake motion. Richard does not appreciate "bases" under his work; even so, he agreed with my idea, asking that I design the same.

After sketching a concept, I sent my sketches to 3M Company, asking if they would provide the isolation details in exchange for appropriate recognition in the exhibit; they agreed to do so. I completed the designs, sending them to Richard who, in turn, provided them to the museum. Their very expert engineer called me to say that he had rejected the proposal once and could not now accept the installation. My interpretation, for which I have no evidence, is that the museum did not want to deal with these very heavy objects being placed on its upper floor. I called Richard who was, to say the least, very upset. Still, following a bit of time, he cooled off.

Various sculptors have experienced unanticipated events, sometimes resulting in injuries and deaths. It is not uncommon for just one of Richard's steel plates to weigh as much as 20 tons (about the same as a dozen cars). Appropriately, he became sensitized to these problems, retaining us for the review of several installations. For example, often, it just wasn't practical to accomplish a proper evaluation of the 100–120-year-old buildings where the sculptures were to be installed. There were no drawings; the floor framing was timber of unknown quality, supported on brick party walls, generally assembled with lime mortar. At a cost far less than a proper review by us, we had timbers and posts installed in the basements below to bolster the existing floors.

For a very large and tall piece to be erected outdoors on a narrow peninsula off of Qatar, there arose the matter of appropriate illumination. Richard suggested that lights could be floated on or under the sea. Taking the other approach, I pointed out that permission need be obtained from the Qatari authorities to place such lights in waters traversed by moderate-sized fishing vessels. I went on to say that we had attempted a below-the-water approach for the nearby Museum of Islamic Art by I. M. Pei, only to find that the quality of the lighting was poor, the cost of the installation and of the maintenance was prohibitive, and that governmental permission was never obtained; for I. M.'s museum, the underwater lighting scheme had been abandoned.

Richard persisted. Then, as he continued speaking, Richard's lovely and wise wife, Clara, kept kicking him under the table: "Richard, no lights in the water!" Before the close of the meeting, Clara's counsel prevailed: there would be no lights in the water.

Assuming that people would pass around and through the sculpture, I queried as to the possible danger from steel heated by the fierce Qatari sun. Of course, SawTeen and I developed details allowing for the horizontal expansion and contraction of the steelwork, while properly resisting the lateral forces from wind and earthquake, all while hiding these details from the viewing public. Recently we were told both that the sliding connections were working perfectly and that Richard had provided assurances that the temperature of the steel would not reach dangerous levels.

Richard, with his sometimes quick temper and his always quick smile, was a real pleasure to work with. At our weekend home in Connecticut, we have a

Collaborating with the sculptor, painter, and printmaker Richard Serra is exhilarating. These are some views of his massive scuplture 7 in Doha, Qatar.

meter-high maquette of his towering 80-feet-high sculpture for Qatar. Unquestionably because I failed to seat it properly on a rock outcropping, it was toppled in a severe windstorm, with each of the plates separating from the others. The real trick was to find a welder both qualified and interested in accomplishing aesthetically acceptable repairs.

BEVERLY PEPPER

For many years, SawTeen and I have enjoyed a delightful relationship with the sculptor/painter Beverly Pepper as well as with her wonderful husband-writer, the late Curtis Bill Pepper. Our personal libraries have books written by Bill and by their Pulitzer Prize winning daughter, Jorie Graham, as well as a variety of books about Beverly. As gifts from Beverly, we own a stainless-steel sculpture now residing in front of our weekend home in Connecticut, and a maquette of one of her *Sentinels*. As well, a delicious print by Beverly greets us at the entrance to our apartment in New York.

Perhaps the most visible of our efforts with Beverly are two sets of 36–39-feet-high *Manhattan Sentinels*, located on a large plaza on the east side of Broadway, in front of the Jacob K. Javits Federal Building in New York's Civic Center (where, ironically, Richard Serra's *Tilted Arc* had stood until its controversial removal in 1989). Beyond the structural work, I carried an active interest in the entire project. For example, Beverly had retained a landscape architect to sort out the trees to be planted around each of the three *Sentinels*. Ginkgo had been selected, an ornamental variety found commonly in New York City; the composition looked great in Beverly's model. Even so, I pointed out that the ginkgo was fast growing to considerable height and breadth and, it seemed to me, would soon hide her work. Properly so, Beverly, relying on her landscape architect, was skeptical of my comment but did reduce the number of trees. Now, with the construction of a new entry pavilion that intrudes into the plaza, in my view, the once-beautiful open space has been severely compromised.

Beverly is not only an incredibly talented sculptor, she is a warm and delightful human being. Now, emerging into her nineties, her current work is simply breathtaking! Recently, SawTeen and I spent a few days at her home and in her studio in Todi Gentile, Italy, and had visited the fabrication works for her current sculptures. As is our way, we were able to provide modest ideas regarding the larger works. In early 2015, in New York at the Marlborough Gallery, she had a beautiful showing of both desktop and larger works. At a dinner party following the opening, she proffered simple words of advice: "Never retire!" Wise words that I fully intend to follow, from a beautiful, energetic, and talented woman (seen with me at the beginning of the chapter).

Working with architects Robert (Bob) Venturi and Denise Scott Brown, we were the structural engineers for the Seattle Art Museum. At the outset, a problem arose in that neither was licensed to practice in the seismic areas of the West Coast. What to do? Well, to assist him in passing the earthquake engineering portion of the architectural examination, every Saturday morning I would drive from my home in New York to Bob's offices in Philadelphia, there to give him lessons in earthquake engineering. It was a great success! Bob, proud of his achievement at passing the examination would say to interested listeners: "I was the only person in the room who passed the examination!" Then, after a respectable pause, he would add: "Of course, I was the only person in the room." For the smaller sculpture at the museum, I designed base-isolation mounts that are invisible to the viewing public.

One of Beverly's sculptures being fabricated in a factory outside of Todi, Italy.

However, for the 48-foot tall, hollow, *Hammering Man*, by the sculptor Jonathan Borofsky, weighing about 13 tons and, being only 7 inches thick, this was not practical. Fabricated in Connecticut, *Hammering Man* was hoisted into position at the construction site in Seattle but collapsed in the process. The erector had fastened the sling around the neck of the sculpture. As one opposed to such things as the burning of Santa Clause candles, I find the hoisting of a sculpture using a noose around the neck to be inhumane, totally inexcusable. In any event, it was returned to Connecticut, repaired and returned to the museum, before being properly hoisted with the sling around the body and under the arms.

Working with Sasaki, Dawson, DeMay and Associates, in the design of sculpture for Binghamton, New York was my first encounter with the late architect-sculptor Masao Kinoshita (1969). Set in a reflecting pool in a sunken plaza, Masao's concept contains intersecting Corten steel H-beams, organized in a flowing but abstract composition. Constructed without stiffeners, perhaps the largest issue was the atmosphere-induced stress from thermal expansion and contraction, particularly at the eccentric connections. Beyond my interest in the complexities of the design, the opportunity to work with Masao was an honor and a pleasure. Once, when he visited our country house on Candlewood Lake, he took the time to sketch a master plan on our contour map, and also accomplished a painting, 4 feet by 8 feet in size, which now adorns a wall in the dining area of our New York apartment.

VEST-POCKET PARKS BY SASAKI, DAWSON, DEMAY AND ASSOCIATES, INC

With Masao Kinoshita, we completed two vest-pocket parks: the first being Greenacre Park (1971), located in midtown Manhattan on a site 60 feet wide by 120 feet deep, with high-rise buildings on three sides and the fourth opening onto East 51st Street (between 2nd and 3rd avenues). Funded by Abby Rockefeller Mauzé, we enjoyed the company of her brother, David Rockefeller, philanthropist and chairman of Chase Manhattan, at most of our project meetings. In one of those meetings, the project's mechanical/electrical engineer commented on the "50-year design life of the project." David seemed to levitate out of his chair, looking down and saying: "Young man, when you are long gone, this park will be serving the peoples of New York." Not the subject of his comment, fortunately, I had set our sights rather higher.

In establishing our fees, I had thought of our role as largely administrative, i.e. meetings, plus a couple of park benches, entry gate, some small walls around planters, that sort of thing. It wasn't until later that I discovered we needed to underpin the edges of all three of the adjacent high-rise buildings, that every square foot of the park was a concrete slab structurally spanning over a drainage system below (of course with protection against entrance by rodents), and that the clearspan entrance gate and trellis work was complex beyond belief. Fortunately, both Masao and David were sympathetic to our plight.

Today, as I take a seat in this tiny park, despite the presence of multi-story buildings on three sides and a bustling street on the fourth, I cannot help but feel comforted by the serenity and the beauty of the place. Of course, it was the leadership of David Rockefeller and the designs by Masao Kinoshita that made possible this bit of heaven in the turmoil of New York City. Needless to say, on a nice day, it is nearly impossible to find an empty space to sit, making the long hours of design more than worthwhile.

Waterfall Garden Park, another project with Masao Kinoshita, seemed simple by comparison. The park was to be constructed at the edge of Pioneer Square, a historic district in downtown Seattle. Alerted by my experiences on Greenacre Park, I elected to tour the site, gaining permission to visit the basement of the adjacent property where, much to my surprise, I discovered that the slab on ground was broken up in large pieces, perhaps 6 feet on a side, and jutting upward as much as 3 feet! Why? I meandered around Pioneer Square, talking with this person and that before encountering an elderly fireman, smoking his pipe just at the door of the firehouse. He held a wealth of historical information, the most important being that this was the site of Henry Yesler's long-abandoned sawmill. There was 40 feet of decomposing sawdust under our site! The park, I realized, would have to be supported on piling.

Another serious matter was that, for seismic considerations, the Seattle Building Code required that each and every stone be anchored into the supporting wall with a horizontal force equal to the weight of that stone. Masao's concept was that boulders, not stones, backed by a concrete retaining wall, would be used to form a 22-feet-high waterfall. The huge horizontal forces stipulated by the Building

Code made this impractical, yet the Building Department remained adamant. It was Masao's idea that solved the problem. Knowing that such walls are commonly constructed in Japan, a highly seismic area, he proposed to bring masons from Japan who would select and place the stones. The Building Department agreed. We had our beautiful waterfall.

Later, with the architect Isoya Yoshida, resident in Japan, and with Sasaki, Dawson, DeMay and Associates, but with Masao at the US helm, we accomplished the structural design for the Japanese Embassy in Washington DC. There are fascinating stories to be told, but for reasons of security, I dare not share them. One event demonstrates something of the need to minimize cross-cultural differences: the mechanical engineer, realizing that the Japanese tend to be heavy smokers, designed the systems with a high level of air filtration. Just in time to save the honor of the chef, it was discovered that the filtration system removed almost all of the cooking aromas, making food almost tasteless. Sometimes one can do too good a job!

With Masao, we completed several other projects. Born in the United States, partially educated in Japan, in his later years he moved permanently to Japan, where he passed in 2003. This complex man, half rooted in one culture and half in another, somehow fit well with me, a country boy from the United States. Even after he abandoned the US for Japan, we remained good friends, corresponding regularly.

Often, there is no great challenge in the design of these small structures. Instead the challenge and the fun comes with the opportunity to work with these fascinating and talented people. For sure, I find these experiences to be most satisfying.

Bottom left Waterfall Garden, by Masao Kinoshita, in Pioneer Square, Seattle, provided challenging issues both above and in the ground.

Below Kinoshita-san executed this 4 foot by 8 foot painting in the front yard of our country home; now it hangs prominently in our New York City apartment.

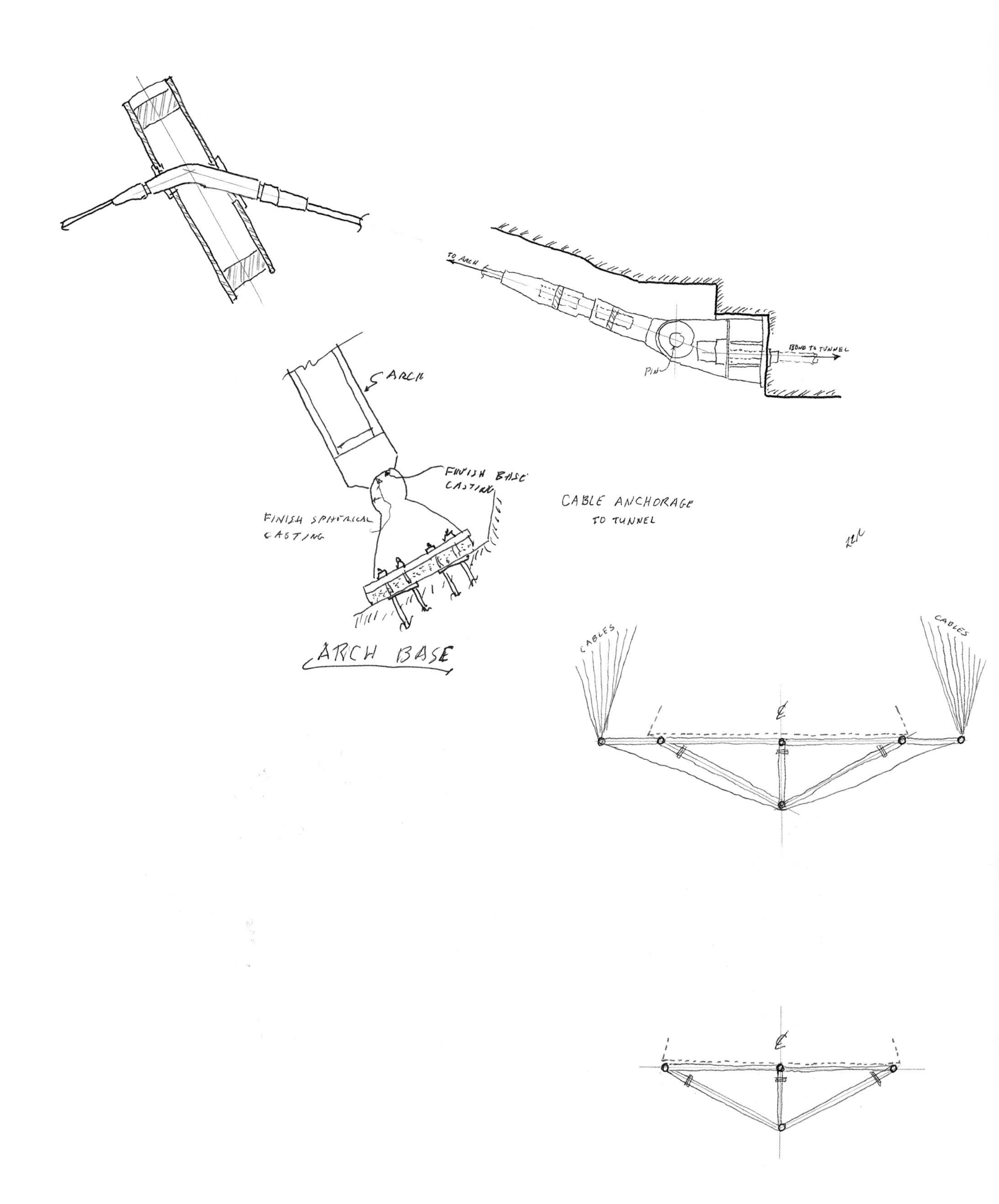
TO ARCH
BOND TO TUNNEL
PIN
ARCH
FINISH BASE CASTING
FINISH SPHERICAL CASTING
ARCH BASE
CABLE ANCHORAGE TO TUNNEL
CABLES
CABLES

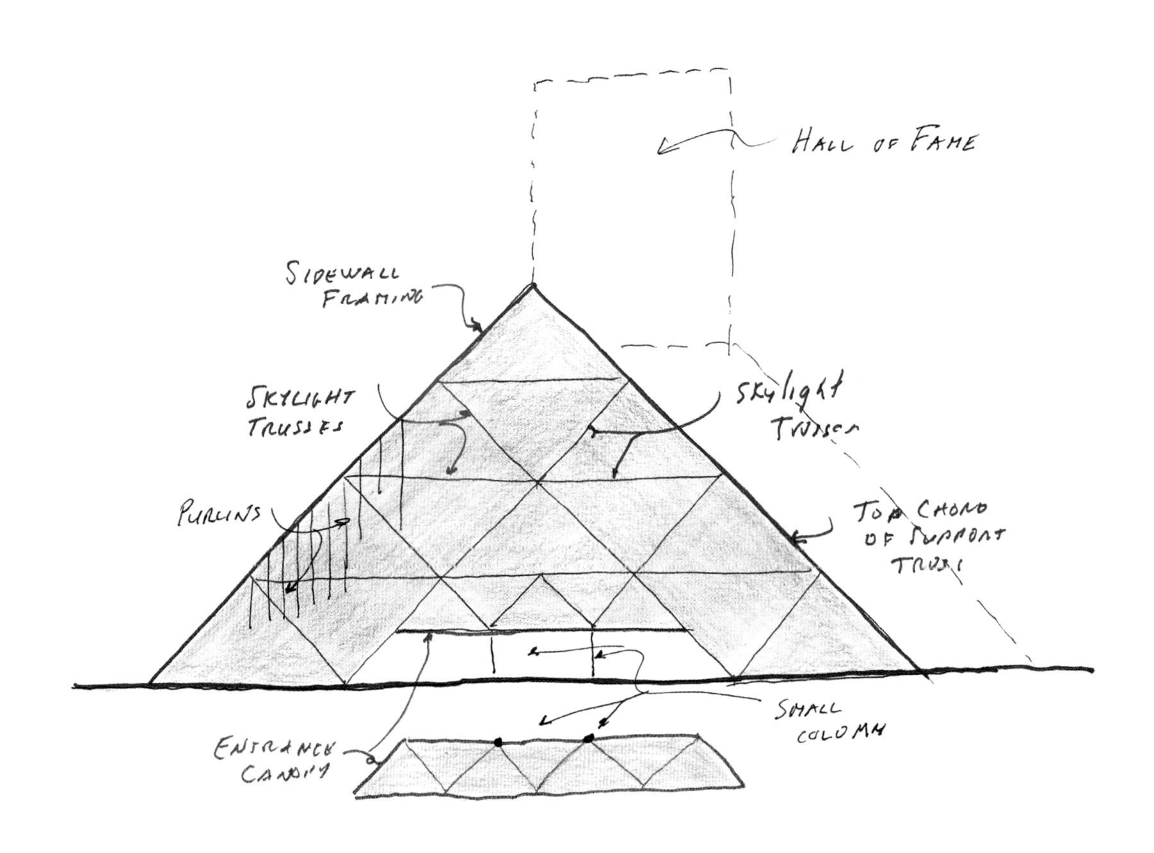
HALL OF FAME
SIDEWALL FRAMING
SKYLIGHT TRUSSES
Skylight Trusses
PURLINS
TOP CHORD OF SUPPORT TRUSS
SMALL COLUMN
ENTRANCE CANOPY

ROCK 'N ROLL

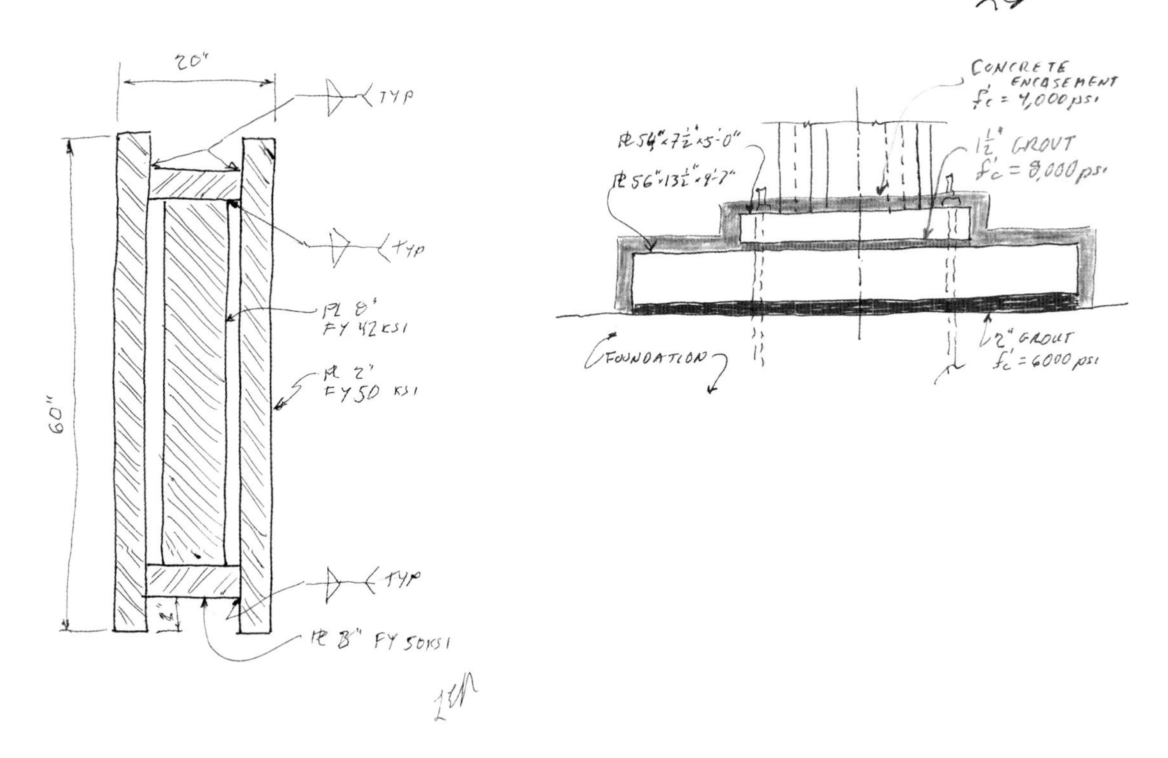
20"
TYP
TYP
PL 8" FY 42 KSI
PL 2" FY 50 KSI
60"
TYP
PL 3" FY 50KSI
CONCRETE ENCASEMENT f'c = 4,000 psi
PL 54"×7½"×5'-0"
PL 56"×13½"×9'-7"
1½" GROUT f'c = 8,000 psi
FOUNDATION
2" GROUT f'c = 6000 psi

Afterword

As my good friend I. M. Pei would say: "There is a beginning that leads to an end." It is where we are now, at the end, where I am struggling for words.

I am a pacifist, with my convictions going far beyond those of war and of fighting. I can be roused to indignation, but not to anger. Yes, I will debate but I do so with both humility and respect. Others say that I am quick to forgive but, more realistically, I don't harbor thoughts in need of forgiving. For all of the above, I am not perfect, nor do I strive for perfection; instead I do my best to do what comes naturally to me: the realization of pacifism and of giving.

What, then, is to be found in a concluding chapter? In this book I have attempted to explore many examples, many adventures, all aimed at providing to architects and engineers something of my life experience in the challenging interfaces between our respective professions. As well, fully appreciating the hard work required to stay just current with contemporary practices and technologies, young practitioners have entered a world where that hard work pays off. But never be satisfied, being current with the latest ideas and approaches is not enough!

Further, the modern world's litigious atmosphere has continued to escalate. It being simpler, let's tackle this last thought first:

The dangers associated with professional errors and omissions surrounds us in all of our work. While my experience with litigation is limited to the providing of advice to lawyers and to one instance of providing a deposition, I have seen so many design professionals struck by the lightning of professional errors and/or omissions as to sharpen my awareness of the dangers before us. Saw Teen continues to preach the gospel of in-depth review and careful attention to words, both oral and written.

In the chapter "Some Views of Architecture and Engineering," the reader will find, "besides being ingenious . . . when sailing in the shark-filled sea of professional indemnity, architects and engineers should be brave in proportion." Brave, yes, but being thoughtfully appreciative and aware of the dangers is even more important. There is little "logic" in all of this. One is able to learn of the many dangers only by reading, by listening to others, by attending lectures and the like, and this learning habit must continue throughout one's professional career. In the chapter "The History of Construction," outrageous examples are given, all with the hope that they will bring a smile to the reader's face, but the dangers are real and the consequences can be severe. Wake up! Stay alert! Failure to do so may find the heedless in a court of law, or with a bank account at zero, or on the street looking for a new job.

The assault of the computer fills our work environment. The teaching of design, analysis, and graphics by computer at the university forms the foundation of most young designers, but not for folks like me, and countless other talented architects and engineers. It is so easy! To analyze a simple beam or to graphically depict a wall intersection, one need only punch a few computer keys to find all of the desired data. As the problem becomes more complex, it is still the same, just a few more keys to punch. One could teach a child to do it. So what is the personal need? It is the ideas generated in one's mind and expressed in one's hand-drawn sketches that, I believe, pave the road to the future.

The majority of young architects and engineers turn to the computer at the very outset of their assignment or project. The wonderful journey of an idea from the brain, out through the neck and shoulders, down through the arms and hands to the fingers grasping a pencil is forgotten by all too many design professionals-in-the-making and, perhaps more importantly, by their professors. The finality of the computer output is so perfect, so ideal, as to be accepted without question.

By watching more experienced and talented designers I learned to sketch first, and to then analyze that which has been sketched. Now, young designers come to me with computer output, streams of numbers, asking my advice on the structure represented by those numbers. Their enthusiasm cannot be faulted and must not be dampened but, all too often, the numbers represent something that they only partially understand, is probably incomplete, and almost certainly is not representative of what, eventually, will be constructed. As the young designer approaches me with his or her reams of numbers, my eyes start to glaze over, my shoulders sag, all leading that designer to understand that I'm not listening. Sometimes the designer arrives with an incomplete sketch of what he or she had analyzed. After being sent away to complete their sketch, they return, chagrined and understanding better the alternative systems that must be considered.

Before concluding, there is the need to iterate the necessity for everyone in our related professions to get out from behind their computer screens, and to become involved in the needs of others who clamor for just a bit of our lives.

Whether it is to press our government, at all levels, to make life better for all of the peoples of the world or volunteerism in a soup kitchen, there are places crying for support, for our participation. In short, each of us must become involved with both time and money in just some of the myriad of "causes" that are vital to the survival of us all, and completely outside of the technical/design facet of our lives.

The thoughts of Albert Einstein deserve our attention: to increase "men's blessings" you need to do more than just master your technology. Concerns for society, both domestic and overseas, should be the foundation of your life. For it is only in this way that your efforts will be "a blessing and not a curse to mankind."

CREDITS

AEDAS, courtesy of Keith Griffiths 48 bottom center
Antandrus, licensed under GNU Free Documentation License Version 1.2 via Wikimedia Commons 16 top left
Chris Arend Photography, courtesy of Anchorage Museum 31 top left
Beyond My Ken, licensed under GNU Free Documentation License Version 1.2 via Wikimedia Commons 39 bottom right
Courtesy of Gunnar Birkerts 18 top right and bottom right, 20 bottom left, 21 bottom left, 21 bottom center, 22 top left and bottom center, 27 top left, 32 top right and bottom center, 39 bottom left, 188, 192, 194 top and bottom, 195, 242 top
William Blake, *Europe a Prophecy*, copy D, object 1, British Museum via Wikimedia Commons 76
Jean-Louis Blondeau/Polaris01681806 92–93
Terri Meyer Boake 67 right
Morgan Catha, licensed under Creative Commons Attribution-Share Alike 3.0 Unported license via Wikimedia Commons 19 top left
William Cho, licensed under Creative Commons Attribution-Share Alike 2.0 Generic license via Wikimedia Commons 35 top center
Courtesy of City of Cupertino 45 left
Courtesy of Cochran Stephenson & Donkervoet/LMN Architects 35 top right
Courtesy of Council on Tall Buildings and Urban Habitat 295
Courtesy of Eric Cumine Associates 24 bottom left
Daderot, released into the public domain via Wikimedia Commons 32 top left
Ellerbe Becket, courtesy of Peter Pran 38 top right
ENR, courtesy of Janice Tuchman 267 bottom
ESCO Co., courtesy of Shizuo Harada 34 bottom right
Estomuybueno, released into the public domain via Wikimedia Commons 24 top left
Courtesy of Fender Katsalidis Architects 48 top center
Courtesy of David Fisher 44 bottom left
Fortunate4now, licensed under Creative Commons CC0 1.0 Universal Public Domain Dedication via Wikimedia Commons 17 bottom right
Courtesy of Fox & Fowle 30 bottom right
Courtesy of Josef Gartner 262
Courtesy of Grad Associates 23 top
Courtesy of Harrisburg Property Services 25 top left
Courtesy of Henderson Land 38 bottom right, 42 top right, 43 bottom right
Gordana Herning, courtesy of Professor Maria Garlock, Princeton University 86
Carol Highsmith, ID highsm.04817, Library of Congress, Prints and Photographs Division 223 top
Courtesy of Katherine Hill 85
HOK, courtesy of William Hellmuth 44 top left
David Holt, licensed under the Creative Commons Attribution-Share Alike 2.0 Generic license via Wikimedia Commons 115 top right
© Timothy Hursley 256 bottom
Courtesy of IBM Corporation 21 top center, 196, 199
Kerun Ip, Courtesy of Pei Partnership Architects 42 bottom left, 290, 293
Stanley Jesudowich, Courtesy of Pei Cobb Freed & Partners 28 top right
Courtesy of Alan Ritchie 22 top center, 24 right, 28 top left, 31 top center, 34 top, 35 top left, 40 top right, 226, 228, 230, 233, 234, 237 bottom, 240, 242 bottom, 244
Gayle Karen, licensed under the Creative Commons Attribution-Share Alike 3.0 Unported license via Wikimedia Commons 91
Kikutake Architects, courtesy of Mutsuko Theresa Smith 37 top, 248, 250, 251
Courtesy of Kohn Pedersen Fox Associates (KPF) 39 bottom center, 40 left, 41 right, 42 top center, 48 bottom right, 49, 298, 300, 308, 310, 312
© Ian Lambot 265
© Bill Lebovich 168
Courtesy of LeMessurier Consultants 210
Courtesy of Lotte Group 46 bottom center
Joe Mabel, licensed under GNU Free Documentation License Version 1.2 via Wikimedia Commons 35 bottom center
Madcoverboy, licensed under Creative Commons Attribution-Share Alike 3.0 Unported license via Wikimedia Commons 25 bottom left (cropped), 28 top center
Courtesy of The Morton H. Meyerson Symphony Center 29 top
Minnaert, licensed by Creative Commons Attribution-Share Alike 3.0 Unported license via Wikimedia Commons 32 top center (cropped)
Courtesy of Mitchell/Giurgola Architects 19 bottom right, 30 top right, 204, 209 top
Courtesy of Mori Building Company 309 top
Kirk Murray, courtesy of University of Iowa 27 bottom left
Courtesy of Nippon Steel Corporation 37 bottom center, 267 top, 268
Courtesy of Guy Nordenson and students 137
© Northstar Gallery 22 top right
Courtesy of Oregon Research Institute 121 bottom left, 139
© Richard Payne 25 top center and top right
Courtesy of Pei Cobb Freed & Partners 35 bottom left, 40 bottom right, 46 bottom right, 255, 258, 266 left, 269, 271

Courtesy of Pei Partnership Architects 37 bottom right, 39 top right, 43 top, 45 top right, 319 top
Courtesy of Beverly Pepper Studio 29 bottom center
The Photographer, licensed under Creative Commons CC0 1.0 Universal Public Domain Dedication via Wikimedia Commons 16 top center, 63
Courtesy of Port Authority of New York and New Jersey 17 left, 106–107, 108, 113 top, 115 top left and bottom, 116 bottom, 122 center and bottom, 125 bottom, 129
Courtesy of Portland Museum of Art 25 bottom right
Researcher Q, licensed under the Creative Commons Attribution-Share Alike 3.0 Unported license via Wikimedia Commons 34 bottom left
Courtesy of the Robertson family 52, 54, 55, 57
Leslie Earl Robertson 22 bottom left, 62 left, 88 bottom, 122 top left and right, 142 top left and bottom, 176 top, 179 top, 182, 183, 187, 200, 229, 232, 243, 245, 260, 261, 266 right, 309 bottom left, 313 bottom, 324–325
Leslie E. Robertson Associates 16 bottom left, 88 top, 109, 113 bottom, 116 top, 118 top, 118 bottom left and right, 121 top left, 125 top, 127, 141, 149, 152, 153, 171, 179 bottom, 181, 214, 216, 238, 264, 273, 302, 305
Fadi Asmar 25 bottom center, 41 bottom left
Sami Matar 16 bottom left, 18 left, 21 top right, 237 top, 239, 316
W. John Pugh 26 left
Courtesy of Rowan Williams Davies and Irwin Inc. (RWDI) 313 top
Seattle Municipal Archives, Item 53332 16 top right, 65
SawTeen See 20 top center, 26 top right, 27 right, 30 left, 36, 39 top center, 41 top left, 70, 73, 79, 142 top right, 145, 147, 150, 151, 156, 190, 202, 206, 223 bottom, 225, 257, 275, 279 top left, 279 bottom, 282 bottom, 287 top, 296, 309 bottom right, 314, 321
© Alexey Sergeev 256 top
Courtesy of Richard Serra 47 bottom left, 319 bottom
Courtesy of Shanghai International Financial Center 48 top right
Courtesy of Shinji Shumeikai cover, 33 top right, 35 bottom right, 46 top, 164, 166, 274, 277, 279 top right, 281, 282 top, 283, 287 bottom left and bottom center
Courtesy of Siso Shaw y Asociados Architectos 23 bottom right
Courtesy of Skidmore, Owings & Merrill (SOM) 42 top left, 42 bottom center and bottom right, 43 bottom center, 47 bottom center, 48 top left
Courtesy of Skilling Helle Christiansen Robertson (SHCR) 64
Robert A.M. Stern Architects, courtesy of Megan McDermott 45 bottom right
Courtesy of Janet Adams Strong 19 top center, 20 bottom center, 21 top left, 22 bottom right, 23 bottom left, 26 bottom right, 31 bottom right, 33 bottom right, 38 left, 39 top left, 222, 288, 323
Courtesy of SWA Landscape Architects 48 bottom left
Swampyank, licensed under Creative Commons Attribution-Share Alike 3.0 Unported license via Wikimedia Commons 31 bottom left
Courtesy of Wendy Talarico 218
Courtesy of Michael Timm 62 top right
Steven Tucker 252
Courtesy of United States Steel Corp. 67 left, 68, 69, 172
University of California, Berkeley, Records of the Department of Architecture, Environmental Design Archives 59
University of Pennsylvania, Annenberg Rare Book & Manuscript Library 83; The Architectural Archives: Mitchell Giurgola Collection 267: 205; Rollin LaFrance Collection 19 bottom center, 20 top right, 21 bottom right
Courtesy of University of Western Ontario 32 bottom left, 121 right, 176 bottom, 177, 178
Courtesy of Venturi Rauch and Scott Brown, Inc. 31 bottom center
© Camilo Jose Vergara 162–163
Matt H. Wade, licensed under Creative Commons Attribution-Share Alike 3.0 Unported license via Wikimedia Commons 28 bottom right
© Paul Warchol 47 top
Wattewyl, licensed under Creative Commons Attribution 3.0 Unported license via Wikimedia Commons 33 left
Ryan Wick, licensed under the Creative Commons Attribution 2.0 Generic license via Wikimedia Commons 209 bottom
Courtesy of Woods Bagot Architects 44 right
Courtesy of World Trade Center, Barcelona 32 bottom right
Courtesy of Adam Wright 19 bottom left
Courtesy of Minoru Yamasaki & Associates 16 bottom center and bottom right, 17 top right, 19 top right, 20 top left and bottom right, 28 bottom left and bottom center, 29 bottom right, 31 top right, 101, 102, 105
Courtesy of Taro Yamasaki 98

SELECTED AWARDS, RECOGNITIONS, and HONORS

The following is a necessarily abbreviated list of awards, recognitions, and honors conferred on Leslie Earl Robertson over the past seven decades.

1952	**Bachelor of Science**, University of California, Berkeley
1957	**Civil Engineer #10591**, State of California (licensed or eligible in all 50 states)
1959	**Structural Engineer #1017**, State of California
1965	**Best Engineering**, IBM Seattle, American Institute of Steel Construction
1967	**Viscoelastic Damper, patent #3,605,953, (with 3M)**
1968	**Shaftwall Partition**, conceived and developed
1969	**Outstanding Concrete Project**, Woodrow Wilson School of Public and International Affairs, American Concrete Institute
1970	**Grand Award for Engineering Excellence**, US Steel Tower, American Council of Engineering Companies, New York Association of Consulting Engineers
1971	**Outstanding Civil Engineering Achievement**, awarded for World Trade Center, American Society of Civil Engineers
1971	**Honor Award for Engineering Excellence for the Federal Reserve Bank of Minneapolis**, American Council of Engineering Companies
1971	**Design in Steel Award for the Federal Reserve Bank of Minneapolis**, American Institute of Steel Construction
1973	**Award for Engineering Excellence for the Federal Reserve Bank of Minneapolis**, American Council of Engineering Companies
1974	**Raymond C. Reese Research Prize**, with Peter W. Chen, American Society of Civil Engineers
1974	**Founding Member & Executive Board**, Wind Engineering Research Council
1974	**Architectural Award of Excellence for the Federal Reserve Bank of Minneapolis**, American Institute of Steel Construction
1975	**National Academy of Engineering**, elected
1976	**Profesor del Curso relativo a Aspectos Generales sobre Proyecto**, Diseño y Construcción, Universidad Nacional Autónoma de México
1981	**Founding Member and Board**, Architects, Designers, Planners for Social Responsibility
1983	**Grand Award for Engineering Excellence**, Brendan Byrne Arena, American Council of Engineering Companies
1983	**First Prize, Engineering Excellence**, Brendan Byrne Arena, New York Association of Consulting Engineers
1984	**Engineering College Council**, Cornell University, appointed (1984–89)
1984	**Commissioner, Commission on Engineering and Technical Systems**, National Research Council, appointed
1985	**Inaugural Richard J. Carroll Memorial Lectureship**, John Hopkins University
1985	**Chairman, Council on Tall Buildings and Urban Habitat**, elected
1986	**Doctor of Engineering**, *honoris causa* and commencement address, Rensselaer Polytechnic Institute
1986	**Engineering College Council**, Cornell University (1986–1987)
1986	**America's Most Creative Engineer Award:** Ove Arup, William LeMessurier, Les Robertson, Lev Zetlin, Rensselaer Polytechnic Institute, American Society of Civil Engineers, *Progressive Architecture*
1987	**Fellow, New York Academy of Sciences**, and Member Committee on Human Rights
1988	**State-of-the-Art of Civil Engineering Award**, American Society of Civil Engineers
1988	**Vice President, Architectural League of New York**, elected
1988	**Doctor of Science**, *honoris causa* and commencement address, The University of Western Ontario
1988	**Building Research Board**, Commission on Engineering and Technical Systems, National Research Council, appointed
1989	**The AIA Institute Honor Award for service to architects**

1989 **Construction's Man of the Year,** *Engineering News-Record*

1989 **Chairman, Panel for Assessment of Wind Engineering Issues in the US,** National Research Council

1989 **US Committee for the Decade of Natural Disaster Reduction,** National Research Council, elected

1989 **Grand Award for Excellence in Engineering,** Bank of China Tower, American Council of Engineering Companies

1989 **Prix d'Excellence, Bank of China Tower,** Association des Ingénieurs, Conseils du Canada

1989 **Best Structure Award, Bank of China Tower,** Structural Engineers Association of Illinois

1990 **Award for Excellence in Engineering, San Jose Convention Center,** New York Association of Consulting Engineers

1991 **Doctor of Engineering,** *honoris causa,* Lehigh University

1991 **John F. Parmer Award,** Structural Engineers Association of Illinois

1991 **Distinguished Engineering Alumnus,** University of California, Berkeley

1991 **James L. Sherard Memorial Lecture,** University of California, Berkeley

1992 **Sway Minimization System for Elevator Cables, patents #5,103,937, #DE 692 19 464.9-06**

1993 **City of New York Certificate of Recognition,** for repairs after the WTC bombing

1993 **The World Trade Center Individual Exceptional Service Medal,** Port Authority of New York and New Jersey

1993 **The Top 25 Newsmakers of ENR,** *Engineering News-Record*

1993 **Dr. Professor Gengo Matsui Prize, Japan, for the outstanding structural engineer in the world**

1993 **Mayor's Award and Medal for Excellence in Science and Technology,** City of New York

1994 **First Grade Architect and Engineer,** #252891, Japan

1994 **Fellow, New York Academy of Sciences**

1994 **Citation of Excellence,** *Engineering News-Record*

1995 **Structural System for Long-Span Structures, patent #11800161** (with Kiyonori Kikutake)

1995 **Life Member,** American Society of Civil Engineers

1995 **National Research Council, Commission on Engineering and Technical Systems, Board on Infrastructure and the Constructed Environment on Feasibility of Applying Blast-Mitigation Technology from Military Facilities to Civilian Buildings**

1995 **WTC Emergency Response Award,** Concrete Industry Board

1996 **Award for Great Contributions to the World of Architecture, Honorary Fellowship and Medal,** Tokyo Society of Architects & Building Engineers

1996 **Honorary Fellow,** Singapore Structural Steel Society, elected

1996 **Board of Directors,** Skyscraper Museum

1996 **Fellow,** MacDowell Colony

1996 **Member,** Architectural Institute of Japan

1996 **First Prize for Engineering Excellence,** Puerta de Europa, Spain, New York Association of Consulting Engineers

1997 **Platinum Award,** Rock'n Roll Hall of Fame & Museum, New York Association of Consulting Engineers

1997 **National Engineering Award of Excellence,** Rock'n Roll Hall of Fame & Museum, American Institute of Steel Construction

1997 **Innovative Design and Excellence in Architecture with Steel,** American Institute of Steel Construction

1998 **Grand Award for Excellence in Engineering Design,** Baltimore Convention Center Expansion, American Council of Engineering Companies

1999 **Honorary Membership** (inaugural recipient) Structural Engineers Association of New York

1999 **Most Innovative Structure, Miho Museum Bridge,** Japan, Structural Engineers Association of Illinois

2000 **The Homer Gage Balcom Award,** For a Lifetime of Excellence in the Structural Engineering of Buildings and Special Structures, The American Society of Civil Engineers Metropolitan Section

2001 **The J. Lloyd Kimbrough Award,** American Institute of Steel Construction

2001 **The Landscape and Design Prize for the Miho Museum Bridge** (I. M. Pei and Leslie E. Robertson), Japan Society of Consulting Engineers

2002 **Outstanding Structure Award for design of the Miho Museum Bridge,** International Association for Bridge and Structural Engineering

2002 **The Frank Howard Distinguished Lecture,** George Washington University

2002 **Henry C. Turner Prize for Innovation in Construction Technology** (inaugural recipient), Turner Construction Company & National Building Museum

2002 **Professor, Department of Civil and Environmental Engineering,** Princeton University

2002 **The Felix Candela Lecture,** Museum of Modern Art, Massachusetts Institute of Technology, Princeton University

2002 **Commencement Speaker,** Pennsylvania State University

2003 **The Gordon Smith Lecture,** Yale University

2003 **Outstanding Projects and Leaders Award for Outstanding Lifetime Achievements in Design,** American Society of Civil Engineers

2003 **Doctor Mechanicarum Artium,** *honoris causa*, University of Notre Dame

2003 **Da Vinci Award,** New Jersey Association of Structural Engineers

2003 **Distinguished Alumnus Award,** College of Engineering, University of California, Berkeley

2004 **The Gold Medal,** Institution of Structural Engineers, United Kingdom

2004 **The Fazlur Rahman Khan Medal** (inaugural recipient), Council on Tall Buildings and Urban Habitat

2004 **Board of Directors,** MacDowell Colony

2004 **The Wall of Inspiration,** Science Center, Cornell University

2004 **Platinum Award for Engineering Excellence,** Espirito Santo Plaza, Miami, American Council of Engineering Companies

2005 **Professor Honoris Causa,** Universitatea de Arhitectura si Urbanism Ion Mincu, Bucuresti

2005 **Honorary Member and Medal,** National Society of Romanian Engineers

2005 **Fellow,** Institution of Structural Engineers, United Kingdom

2005 **Fellow and Inaugural Y.K. Cheung Lecture,** Hong Kong University

2005 **Platinum Award for Engineering Excellence,** National Constitution, American Council of Engineering Companies

2006 **Distinguished Member,** American Society of Civil Engineers

2006 **Engineer of the Year,** Society of Indo-American Engineers and Architects

2006 **Fellow, CEng FIEI Engineers,** Ireland

2007 **Committee on Human Rights,** New York Academy of Sciences

2007 **Honorary and Advisory Board Member,** International Center for Sustainability, Accountability and Eco-Affordability for Large Structures

2007 **Silver Award for Engineering Excellence,** AIG Tower and Footbridge, American Council of Engineering Companies

2008 **National Honor Member,** Chi Epsilon National

2008 **The Best Tall Building in the World,** Shanghai World Financial Center, Council on Tall Buildings and Urban Habitat

2008 **Platinum Award for Engineering Excellence,** Museum of Islamic Arts, American Council of Engineering Companies

2009 **Lecture,** Technical University of Civil Engineering of Budapest

2009 **Diamond Award for Engineering Excellence,** Shanghai World Financial Center, American Council of Engineering Companies

2010 **Newmark Distinguished Lecture: Creating**

Progress in the World of Design, Chi Epsilon National

2011 **International Award of Merit in Structural Engineering**, International Association for Bridge and Structural Engineering

2011 **Inaugural Nucor Distinguished Lecture**, Virginia Polytechnic Institute and State University

2012 **The 2011 John Fritz Medal for outstanding scientific or industrial achievement**, American Association of Engineering Societies

2012 **Inaugural Member of Academy of Distinguished Alumni**, University of California, Berkeley

2012 **Premio de Ingeniería Civil**, Fundación José Entrecanales Ibarra, Spain

2014 **Fellow**, MacDowell Colony

2014 **Excellence in Structural Engineering Award**, Richard Serra's 7, National Council of Structural Engineers Association

2015 **Structural Engineers World Congress Roland L. Sharpe Medal**